CUMULATIVE TRAUMA DISORDERS

Current Issues and Ergonomic Solutions:
A Systems Approach

Kathryn G. Parker
Harold R. Imbus

CRC Press
Taylor & Francis Group
Boca Raton London New York

CRC Press is an imprint of the
Taylor & Francis Group, an **informa** business

First published 1992 by Lewis Publishers

Published 2019 by CRC Press
Taylor & Francis Group
6000 Broken Sound Parkway NW, Suite 300
Boca Raton, FL 33487-2742

©1992 by Taylor & Francis Group, LLC
CRC Press is an imprint of Taylor & Francis Group, an Informa business

First issued in paperback 2019

No claim to original U.S. Government works

ISBN-13: 978-0-367-45026-7 (pbk)
ISBN-13: 978-0-87371-322-1 (hbk)

Visit the Taylor & Francis Web site at
http://www.taylorandfrancis.com

and the CRC Press Web site at
http://www.crcpress.com

Library of Congress Cataloging-in-Publication Data

Parker, Kathryn G.
Cumulative trauma disorders : current issues and ergonomic solutions:
   a systems approach / Kathryn G. Parker, Harold R. Imbus.
      p. cm.
   Includes bibliographical references and index.
   ISBN 0-87371-322-2
   1. Overuse injuries. 2. Overuse injuries—Prevention. I. Imbus,
   Harold R.
[DNLM: 1. Human Engineering—methods. 2. Occupational Dis-
   eases—etiology. 3. Occupational Diseases—prevention & control.
   4. Stress—etiology. 5. Stress—prevention & control. WA 440 P241c]
RD97.5.P37   1992
616.98´03—dc20
DNLM/DLC
for Library of Congress   92-20620
                                                                    CIP

PRINCIPLE AUTHORS:

Kathryn G. Parker, MS
Consultant in Ergonomics

Harold R. Imbus, MD
President, Health & Hygiene, Inc.

CONTRIBUTING AUTHORS:

R. Keith Butterfield, MS
Peter G. Johnson, MD
A. Ritchie Lewis, MD
Sheila H. Mullins
Anthony P. Wright

PREFACE

A number of books have been written on ergonomic assessment and prevention of cumulative trauma disorders. There have also been many articles in the medical literature pertaining to these disorders. However, seldom is the subject treated in a comprehensive and integrated fashion. We attempt to do so in this book.

Ergonomics is the mainstay in preventing cumulative trauma disorders. Early detection is important for the treatment of cases, and ergonomic programs will fail to achieve their full potential without adequate early detection. However, further ergonomic follow up is vital in controlling the factors that cause the case in the first place.

Cumulative trauma cases often start as simple muscle pain and fatigue. Frequently they are complicated by overzealous treatment, including unnecessary surgical care. Therefore, to control cumulative trauma disorders, there must be a continuous interaction between ergonomists, medical providers within the plant setting, and medical specialists outside the plant setting.

This book, while presenting the subjects in a traditional, categorical approach to identifying ergonomics and medicine, tries to integrate these approaches to assist management, ergonomists, and medical care providers in acheiving optimum results.

Although technical, this book is written in a manner that we hope will be readily accessible to most of those normally responsible for dealing with these problems.

EDITORS

Kathy Parker is an independent consultant in ergonomics, specializing in soft tissue fatigue and cumulative trauma, and is based in Chapel Hill, North Carolina. She has a dual background. She is a licensed physical therapist with a Bachelor of Science degree from the University of North Carolina at Chapel Hill in 1984. In 1986, she earned a Master of Science degree in Industrial Engineering from North Carolina State University with a study emphasis in occupational biomechanics.

Her experience includes ergonomics consulting, training, and managing ergonomics programs for a variety of manufacturing and service companies. She has a special interest in the textile and apparel industry, for which she has provided a variety of programs.

Ms. Parker was previously employed as the Director of Ergonomics for Health and Hygiene, Inc. in Greensboro, North Carolina, where she initiated and developed their ergonomics consulting services.

Dr. Harold R. Imbus is the chairman of Health & Hygiene, Inc. He received his M.D. and Sc.D. from the University of Cincinnati. Dr. Imbus has practiced occupational medicine for three decades and is an international expert on medical surveillance, respiratory testing, and worker compensation issues of byssinosis (cotton dust lung disease). He has published numerous scientific articles on toxic substances and human health, including studies on toxic metals (boron, cadmium, chromium, nickel), formaldehyde, hydrocarbon solvents, and cotton dust. His publications include a chapter on clinical occupational medicine published in Zenz's classic text, *Occupational Medicine*. Dr. Imbus served 12 years as Medical Director for Burlington Industries and Deputy Medical Director for Kennedy Space Center.

Dr. Imbus has over 30 years experience dealing with cumulative trauma disorders as an occupational physician. He has seen a rapid increase of these disorders in the last decade and has been extensively involved with them in recent years. Dr. Imbus has consulted with many large companies pertaining to medical and ergonomic prevention and management of these disorders.

TABLE OF CONTENTS

Introduction

Nerve entrapment syndromes, myalgia, tendonitis, tenosynovitis, and some back injuries are conditions collectively known as cumulative trauma disorders (CTDs). Carpal tunnel syndrome and other nerve entrapment syndromes are typically the end result of musculoskeletal disorders associated with overuse/overexertion. Workplace factors associated with the development of CTDs include

1. Repetition and inadequate work/rest scheduling
2. Force requirements
3. Awkward and extreme postures
4. Static positioning or holding

These disorders are increasingly becoming a concern to occupational health and safety professionals who must deal with them, as well as for the manufacturing plants who employ people affected by these disorders.

The Bureau of Labor has shown an increasing number of reported cases of various CTDs. In 1989, the incidence of CTD cases jumped by 32,000, from 115,000 in 1988 to 147,000 in 1989, accounting for most of the increase in job-related illnesses in the U.S.[1] Job-related CTDs have more than tripled between 1983 and 1987, with some officials predicting that the annual number of untreated or undetected cases may reach 250,000.

Workers' compensation costs are escalating. Thirty to forty percent of current workers' compensation claims are related to musculoskeletal disorders. The amount paid in workers' compensation for lower back injuries more than doubled in 6 years (from $14 billion in 1979 to $30 billion in 1985); two-thirds of all lower back injuries in the workplace are estimated to be CTDs. Some predict that by the turn of the century, 50¢ for every dollar spent in medical costs will be related to CT. The medical community disagrees on whether many CTDs require surgical intervention, which often results in inefficient care and recovery with less than optimal results.

In the 1990s, as manufacturing competition heightens, profit margins shrink, and medical cost containment becomes an economic necessity, it is vital that CTDs be placed into the proper perspective so they can be controlled with a minimum of employee suffering, productivity loss, and expense.

It is worthwhile to discuss some of the trends that have brought CTDs into the limelight in the late 1980s. These trends can be separated into three primary areas of impact:

1. Workplace
2. Workforce
3. Social/medical

All of these trends became acutely present in the later 1980s and remain as issues that must be dealt with in the 1990s.

WORKPLACE TRENDS

Absenteeism

Related to the various social stressors now placed on the average family, absenteeism is an increasingly problematic area for employers to address. Sick children, outside interests, and disputes with supervisors and/or company policy all contribute to absenteeism problems. When a work force cannot be depended upon for daily staffing expectations, those employees that do report to work face the conflicts of increased work demands to help the company meet production expectations while "taking up the slack" of the absent employees.

Specialization

As production lines came into vogue during the latter half of the 1900s, job specialization resulted. Industrial engineering and work study techniques afforded management the opportunity to break the overall production tasks into smaller and smaller elements. A single employee was then assigned very specific combinations of work elements that contributed to the overall efficiency of "the product flow". However, as jobs became more specialized, requiring employees to perform single, discreet tasks, jobs also became more "concentrated", requiring the employee to utilize singular motions and efforts repeatedly over the course of the shift. This lack of task variety, as a result of job specialization, contributes to the potential of overuse in a single muscle group.

Consumerism

Daily schedules have become busier and busier for the typical American. For the sake of "time management", consumer needs have dramatically changed. Manufacturers now offer products that provide the convenience and efficiency that the busy population demands. For example, the red meat and poultry industry responded to the need for convenience in the kitchen by providing "boneless" products. This poses a contrast to the butcher in the corner grocery store of yesteryear who portioned the meat as each shopper desired. This "portioning" and deboning now occurs on a production line. Knives and scissors are repetitively manipulated with the hand and arm to remove the irregularly shaped bones in the various meat products.

Another form of consumerism is reflected in fashion, which specifically impacts the apparel industry. For example, every time the fashion designers determine that pleats are "in", snaps are preferred to buttons, or heavy denim is preferred to stretch knits, the sewing machine operators are significantly affected. They must cope with the new tasks required to hold a pleat as it is sewn, to push snaps together with the fingertips, or to apply more force with the fingertips to handle the heavier and bulkier denim material. In high-style industries, such as sports clothing, these changes can occur each season. It is often very difficult for equipment manufacturers and engineers to automate some of the simpler processes, as they literally may be "here today and gone tomorrow" as a result of fashion whim.

Office Automation

The 1980s quickly became the era of "information". The use of video display terminals (VDTs) is now commonplace for all levels of employees. Office support and clerical staff may spend the majority of their workday sitting in a chair at a desk working with a VDT, reading from documents, listening to tapes for transcription, operating word processing programs, using a keyboard, and editing work by viewing a screen. In order-taking, distribution, telemarketing, and switchboard operations, the seated computer workstation predominates.

VDTs and associated furniture have been associated with various musculoskeletal problems in the hands, arms, neck, back, and legs. Chronic headaches, eye strain, and potential reproductive damage from low frequency radiation have also been linked to VDTs.

Musculoskeletal problems can be tied to the design, layout, and use of various office furniture components. Depending on how the operator adjusts a chair's "ergonomic" features, the muscles can either be relaxed or fatigued. The keyboard is now used continually. Erasures, corrections, and carriage return tasks (previously involved in the use of typewriters) are replaced with constant keystroking with the fingers and static body posturing to position the hands. Lighting and screen orientation affect the eyes and neck, and the general workspace and terminal orientation has been associated with radiation concerns.

Legally, these issues are being addressed by movements in many states to enact laws to protect employees from inadequate workplace furniture, layout, and potential radiation concerns.

Standards were established by ANSI (American National Standards Institute) in the 1970s to provide basic guidelines regarding computer keyboard, screen, lighting, and furniture parameters that are acceptable from a human factor design perspective.

A San Francisco ordinance was enacted in January, 1991 in an attempt to govern VDT use in that area. It provoked strong opposition in the business community and has been threatened by costly legal challenge. The law mandates that

- VDT workstations and chairs be adjusted, meeting certain minimum standards
- VDT users get a 15 min alternate work break every 2 h of VDT use
- Training shall be provided for VDT users about health and safety concerns and repeated on an annual basis
- New employees must be trained within 30 d
- The director of public health can inspect companies for compliance without notice and issue fines up to $500.00
- Employees may not be disciplined for demanding their rights under this ordinance
- Research is to be funded to investigate reproductive and vision issues related to VDT use

WORKFORCE TRENDS

Women in the Workforce

According to the United States Bureau of Labor Statistics, females have recently entered the workforce in record numbers.[2] The number of female workers made up one third (33%) of the total workforce in 1960. In 1987, females comprised almost one half (45%) of the workforce. Working females numbered 57 million in 1990, representing 58% of the workforce.

By introducing more females into the workforce, an interesting dilemma surfaces. Many manufacturing plants were built in the last 15 to 30 years, and their design reflects the male-dominated working population of yesteryear. Therefore, equipment, tools, and general plant layouts were designed for the "average Joe". The average, or fiftieth percentile, male is 69 in. in height.[3] The tallest (ninety fifth percentile) female measures 67.8 in., which still falls short of the "average Joe". The "average Susie", or fiftieth percentile female, at 63.6 in. in height, is dramatically affected by existing workplace reach distances and work heights.

Tool design improvements have focused more on faster, more efficient *tool* performance, and less on the changing work populations and dimensions required for safe and comfortable handling. This basic design conflict exists as engineering concerns are typically product-oriented, with improvements concentrating less effort in accommodating the workforce. "It's always been done that way" is heard too frequently from those manufacturers who are reluctant to retool and invest in the research and development for minor modifications of an established tool design. The conflict between equipment and machine manufacturers' interests, and the ability of the workforce to comfortably interact with their products, poses yet another dilemma for the employer attempting to buy "state-of-the-art" equipment.

Demographics

In the "melting pot" of the U.S., employers need to pay specific attention to the design criteria of the specific work populations. The science of anthropometrics details the differences in the physical attributes of varying populations, nationalities, and groups of people. A plant that is established close to the Mexican border needs to provide equipment that accommodates the Hispanic population that will probably work there. It is questionable whether population differences have been considered in the varying adjustability features offered by equipment manufacturers.

An example of this was observed in a joint Japanese-American manufacturing venture. All workstations installed in their U.S. plant were shipped from Japan. Not surprisingly, a rash of low back and neck problems subsequently arose among the American workers. The workstations, perfectly acceptable to the Japanese workers, were too low in height for the Americans, even when moved to the uppermost height adjustment.

Age

In general, the population is aging. Medical technology allows us to live longer, and therefore work longer. Some estimates cite that by the year 2050, one third of the U.S. population will be at least 55 years old.[4] Anthropometric studies indicate that, in general, muscle strength decreases with age, which can affect the force requirements of a particular job. Also, the body's musculoskeletal tissues age over time, producing conditions such as osteoarthritis which can affect an employee's ability to tolerate otherwise "normal" work stresses. Physical performance also declines with normal aging, which affect joint mobility and muscle strength and flexibility. Employees that are 55 years old today may be performing job tasks with the same physical requirements they were performing at age 20.[4]

Physical Ability

Although our population has been living in an era of wellness and fitness for the past several years, technological improvements in the home have led to a generation of "couch potatoes". The younger generations are no longer expected to perform the physical chores and tasks of the past decades to help keep the household going. More and more time is afforded to television, video games, recreation, and pleasure. Therefore, new workforce generations will be less physically capable to handle jobs that require muscular strength and endurance. Furthermore, it is highly likely that a teenager in the 1980s never worked in an 8-hours-per-day job until he/she graduated from high school. The teenager is less conditioned to handle the physical requirements of full-time production work and may actually be more susceptible to CTDs.

SOCIAL TRENDS

Medical Community

Traditional medical models have supported surgery, medication, and prolonged rest for chronic musculoskeletal conditions. As the rehabilitation disciplines evolve, these models are being challenged in order to improve employee function and reduce disability resulting from musculoskeletal disorders. Family practitioners in a small community face a disconcerting dilemma when an employee requests 2 weeks away from the job for a simple muscle strain. Conflicts arise between pleasing employees, family, employer, maintaining a positive stature in the community, and adhering to an evolving medical model for the treatment of such disorders as tendonitis, myalgia, or disc disease. Seemingly simple decisions about recuperative time and return-to-work-status become more and more difficult for today's "plant doctor".

In the U.S., CTDs seem to have become synonymous with carpal tunnel syndrome in the minds of many physicians and the lay public. Although there are no data as to the percentage of these cases which become diagnosed as carpal tunnel syndrome, this book's co-author, an occupational physician, has visited manufacturing and processing plants throughout the country and has seen many instances of carpal tunnel surgery "epidemics". For example, in one plant over a 2-year period, of 111 CTD cases which were referred to physicians for care, 18 (16%) had surgery for carpal tunnel syndrome. Review of the records of these 18 cases revealed the following facts.

- Most had symptoms of shoulder and neck problems and often forearm problems either before or during the advent of hand symptoms.
- Usually, cases of numbness or paraesthesia were ill-defined and involved; were not typical of carpal tunnel syndrome.
- Electrodiagnostic studies were either essentially normal or very slightly abnormal. There was no standard protocol followed in the electrodiagnostic studies. For example, in some cases only the affected

extremity was done, in other cases only sensory nerve conductions were done.

- Most of these cases were operated on by well-qualified hand surgeons, who are respected by their medical peers.

Outcomes of these cases were as follows: (1) two patients were able to resume their regular work which was repetitive, but not forceful; (2) one of these is continuing to have significant pain and swelling of the hand and probably will have to be taken off the job; (3) three cases have made a satisfactory rehabilitation, but to entirely different jobs not involving repetitive motion; (4) three cases have returned to work, but are on very light duty with no real end in sight; and (5) the remaining cases have either never returned to work, or have returned for short periods and are off with further disability or repeat operations. One might state that four out of eighteen, 22%, had a satisfactory result, but in three cases, alternate work was necessary to achieve a favorable result. One can only wonder if providing ergonomic solutions combined with focused and intensive conservative medical treatment before surgery might have provided similar, if not better, results.

Having toured many facilities, examined and reviewed records of hundreds of pre- and postsurgery employees, it appears that carpal tunnel syndrome is tremendously over-diagnosed in the U.S. and even when correctly diagnosed, surgery is often done early without adequate trial of other therapies, including ergonomic intervention, alternative work, or conservative medical treatment.

Workers' Compensation and OSHA

Over the last 10 years, there has been an upsurge in the number of cases of CTD. These conditions have been identified by several names over the years, but they share the characteristic of lacking a specific initiating event. This has required a change in the workers' compensation system. The system was designed to deal with the acute injury, and there was an agreement between the employer and the employee that a specific consequence came from a particular occurrence. An exception to this "ouch principle" was made in the 1930s when low back pain was administratively defined as an injury, even if a particular episode could not be identified as the trigger.

The workers' compensation laws in most states have now been reinterpreted to include illnesses as well as injuries. These illnesses lack the identifiable starting point, so causality is often disputed. As well, conditions such as carpal tunnel syndrome, tennis elbow, bicipital tendonitis, etc., are not uniquely occupational events. These illnesses commonly occur as the result of daily activities unrelated to work. However, epidemiological studies have established correlations between certain work situations and an excess occurrence of these diagnoses. There has been a transfer of these cases from the employee health benefits area to the workers' compensation arena, and the occupational medical provider is now required to be proficient in diagnosing and treating these conditions.

Treating these conditions in an occupational environment involves fitting the employee into a work environment which will, at a minimum, not make the condition worse, and ideally, will promote healing. This makes medical sense, but there is a legal imperative as well. Under the OSHA General Duty clause, the employer must provide a safe working environment, and in the area of CTDs, the employer will require medical guidance to meet their obligations. OSHA regulatory activity linking ergonomic issues to the General Duty clause dramatically increased in the 1980s. Once the workers' compensation rule changes and increased OSHA regulatory activity made CTD diagnoses an occupational concern, the number of cases increased exponentially. This trend will continue as OSHA works to create an ergonomics standard by the mid 1990s.

Therefore, the employer is further interested in the medical management of these cases, as he or she is liable for penalties for "medical mismanagement". Medical mismanagement can be demonstrated if the employees are dissatisfied with the medical treatment provided at the direction of the employer and if the employer is aware of this dissatisfaction. In treating CTDs, there is a complex interaction of medical, legal, and economic forces which require the health care provider to know the particular CTD diagnosis, to appreciate their natural history, and to be able to interact with the employee to reduce the severity of the condition and to prevent its recurrence.

LEGAL SYSTEMS

In general, litigation is advocated and promoted when we do not feel we have been appropriately treated. Virgil R. May, Ph.D., studied the effects of many variables, including litigation and attorney influence, on workers' compensation claimants' length of disability and overall lifting capability.[5] Interestingly, the type of diagnosis (back, neck, upper extremity, or lower extremity musculoskeletal injury or illness) did not affect the patient's time on disability. However, surgery significantly prolonged the length of disability, as did attorney influence. May states, "When an attorney was involved with a work-related injured person, medical costs and the time it took for resolution of damages significantly increased over those patients who chose not to seek legal counsel."[5]

PURPOSE OF BOOK

All of these issues, although briefly stated, present complexities within themselves; the picture becomes more distorted and more difficult to manage when many of these "trends" exist within a single community and/or place of employment. Although challenging, we can choose to either ignore these trends or adapt to them. Our approach to these trends will directly affect, either positively or negatively, the future productivity, safety, and health of our workforce, especially as it relates to CTDs.

This book intends to offer a discussion of current knowledge and experience in the etiology of CTDs and associated workplace factors, and suggests an approach for control of CTDs in a working environment. This control program is designed as an in-house management system to deal with CTDs. It is designed from a basic occupational health approach in managing illness, which suggests two levels of intervention: primary and secondary.

Primary Prevention

Primary prevention involves the prevention of an occupational disease/illness by: (1) identifying potential hazards and causative agents in the workplace, and (2) addressing them through engineering or administrative controls.

The science of ergonomics offers primary prevention for CTDs. Ergonomics is defined by the ANSI 2794.1-1972 as "a multi-disciplinary activity dealing with the interactions between man and his total working environment, plus such traditional and environmental aspects as atmosphere, heat, light, and sun, as well as of tools and equipment of the workplace." Or more simply stated, industrial ergonomics involves the design of work to go *with* the body, not against it. Therefore, this will involve structuring work tasks, methods, workplace layout, and equipment design to accommodate the work population's strength, endurance, and dimensional capabilities.

Secondary Prevention

Realizing that 100% removal of the potential for CTDs is difficult, a case management model should be established to identify, treat, and manage these illnesses early in their development when they are most easily treated and resolved. Secondary prevention attempts to minimize losses in productivity, disability, and cost associated with those CTDs that may develop in spite of the most ergonomically designed workplace.

Therefore, this book will discuss a rationale for an approach to CTDs in the workplace, including the identification of workplace stressors and establishing engineering, administrative, and medical controls. This book intends to apply the authors' experiences to keep the rationale and methods of this control system "realistic" for the reader. This is not a technical book on ergonomic measurement techniques, design theory, or a recounting of epidemiology or research findings. It is hoped that the reader will find it both thought provoking and practical.

REFERENCES

1.   "Repetitive Motion Injuries Cause Gain in Job Illnesses," *Greensboro News & Record*, November 15, 1990.
2.   U.S. Bureau of Labor Statistics, *Employment and Earnings*, 1987.
3.   Drillis, R. and Contini, R., *Body Segment Parameters*, B. P. 174-942, technical report #1166.03, School of Engineering and Science, New York University, New York, 1966.
4.   "Diverse Workforce Posing New Challenges in Designing Ergonomic Workplace," *Occupational Safety and Health Reporter*, The Bureau of National Affairs, 0095-3237/90, p.995.
5.   May, V. R., "Analyses of Work Functioning Data From a Work Capacity Evaluation Program," Paper presented at the Virginia Orthopaedic Association Annual Spring Conference (Williamsburg, VA: May 1, 1987).

PART I
CAUSATION OF CUMULATIVE
TRAUMA DISORDERS

Chapter 1
Body Tissues and Fatigue

BASIC ANATOMY OF THE UPPER QUARTER

CTDs may affect many parts of the body. However, the major emphasis recently has been on upper extremity conditions. When considering these afflictions, the whole upper quarter of the body — the upper back, neck, shoulder, arm, forearm, wrist, and hands should be included. The shoulder girdle, unlike the pelvic girdle, is not connected directly to the spine. This enables the upper extremity to exhibit a wide variety of motions such as reaching forward or overhead. On the other hand, this structural "latitude" can promote conditions affecting the upper back, neck, and shoulders that are also likely to affect the arm, forearm, wrist, and even the hands.

The nerves that innervate the muscles and provide sensation for the upper extremity originate in the neck. Therefore, impingement on these nerves in the neck can create problems in the arm or hands. The vascular supply to the upper extremity comes from arteries and veins that pass into the arm between the clavicle and first rib. Therefore, muscle problems tending to compress this space can have a profound effect upon circulation in the arm (Figure 1).[1]

When considering CTDs of the upper extremity, one must even consider the mid- and lower back. A painful problem in the low back or one that causes listing of the body can result in abnormal positions of the upper body and cause undue stress upon the trapezius muscles and those of the shoulder and arm (Figure 2).[2]

Although the upper quarter is a complex integrated system of muscles, bones, nerves, arteries, and veins, we will discuss them under the headings musculoskeletal system, nervous system, and circulatory system.

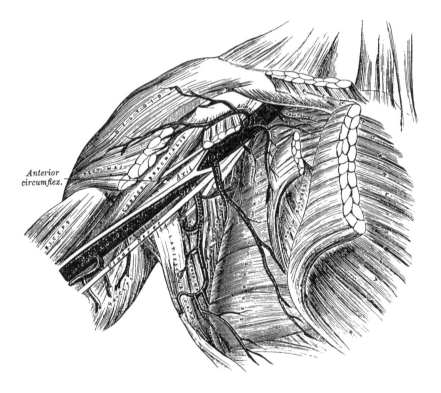

Anterior circumflex.

FIGURE 1. The thoracic outlet; bone, muscles, blood vessels, and nerves are located in close proximity.

Musculoskeletal System

One must consider the vertebral column, especially the cervical vertebra, in the neck when evaluating upper extremity disorders (Figure 3).[3] Altogether there are 33 vertebrae. The 7 cervical vertebrae have smaller bodies than the others. The first two articulate together with the skull in such a way as to allow excellent rotary motion as well as flexion, extension, and lateral bending.

The shoulder girdle is formed by the scapulae (shoulder blades) and clavicles (collarbones). The clavicles articulate with the sternum (breastbone). These are the only bony articulations the shoulder girdle has with the rest of the body (Figure 4).[4] In the back, the scapulae have a wide gap between each other and are connected to the trunk by muscles only (Figure 3). This is in marked contrast to the pelvic girdle which forms a complete ring and connects directly to the spine (Figure 3). The pelvis is massive and comparatively rigid compared to the light and mobile shoulder girdle. This feature of the shoulder girdle provides us with a great advantage in reaching far and wide, but also has the disadvantage that stressful situations which involve the muscles of the

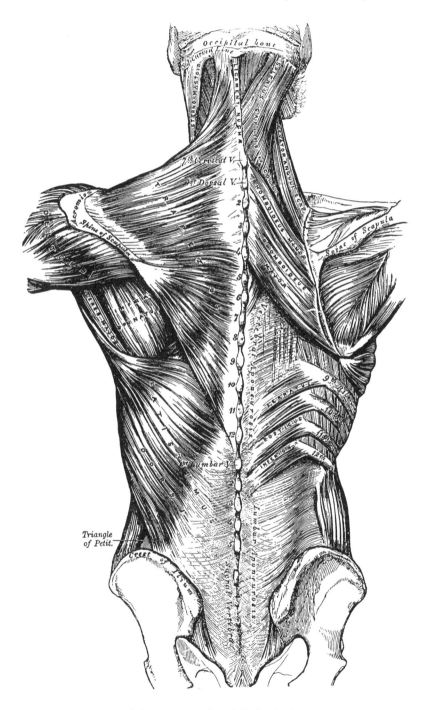

FIGURE 2. Muscles of the upper and middle back: the trapezius, rhomboid, and latissimus dorsi muscles. From Gray, H., *Anatomy, Descriptive and Surgical*, (New York: Bounty Books, 1977), p. 338. (With permission.)

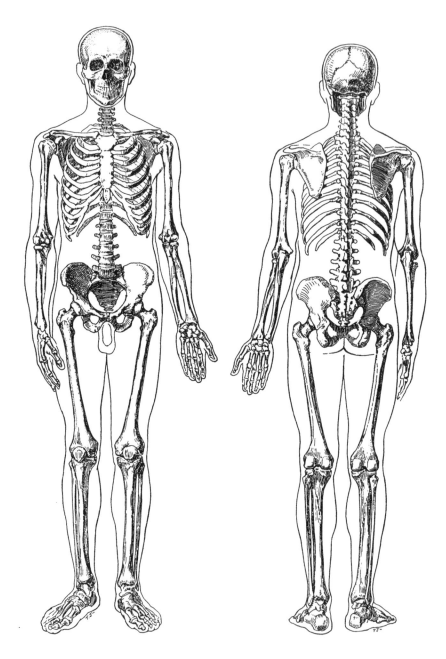

FIGURE 3. The skeleton, front and rear views. From Gray, H., *Anatomy of the Human Body*, (Philadelphia: Lea & Febiger, 1948), p. 74. (With permission.)

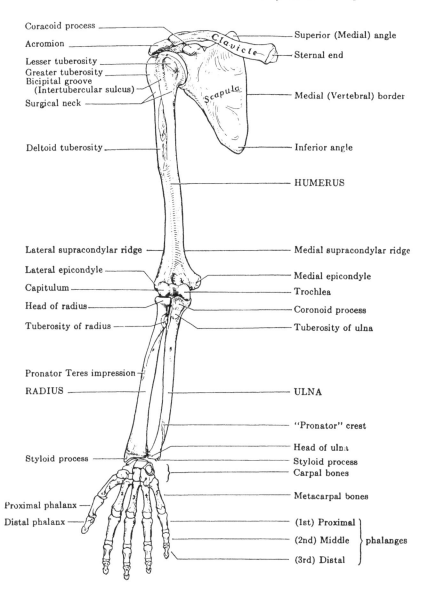

FIGURE 4. Bones of the upper limb, from the front. Grant, J. C. B., *An Atlas of Anatomy*, (Baltimore: Williams & Wilkins, 1947), p. 2. (With permission.)

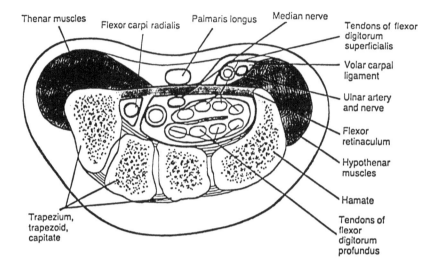

Thenar muscles
Flexor carpi radialis
Palmaris longus
Median nerve
Tendons of flexor digitorum superficialis
Volar carpal ligament
Ulnar artery and nerve
Flexor retinaculum
Hypothenar muscles
Hamate
Trapezium, trapezoid, capitate
Tendons of flexor digitorum profundus

FIGURE 5. The carpal tunnel, formed by the carpal bones, humerus, radius, and the carpal ligament.

region can result in profound changes affecting the entire upper extremity.

The scapula is a flat triangular bone with which the arm articulates (Figures 3 and 4). The acromion of the scapula is the highest part of the shoulder and articulates with the clavicle at the acromio-clavicular joint. The scapula has a cavity which articulates with the head of the humerus which is the long bone of the upper arm. The humerus narrows to a shaft in its mid-portion and then gradually widens at its distal (far) end to the medial (toward the midline of the body) epicondyle and the lateral (away from the midline) epicondyle. The lower end of the humerus articulates with the radius on the lateral side and with the ulna on the medial side.

The radius and ulna make up the bones of the forearm. The ulna, which articulates with the humerus in a hinge fashion, has a large process called the olecranon which is the posterior tip (back part) of the elbow. The radius moves in elbow flexion (upward bending), elbow extension (straightening or backward bending), and forearm rotation. The radius articulates with the humerus and ulna at the head of the radius to effect these movements.

The radius and ulna join with the eight carpal bones at the wrist. The carpal bones form the posterior, medial, and lateral borders of the carpal tunnel (Figure 5).[5] The carpal bones articulate with the five metacarpal bones. These articulate with the phalanges, or finger bones. There are two phalanges in the thumb and three phalanges in the other fingers.

There are numerous muscles of the neck and shoulder girdle, but of particular interest is the trapezius which is a flat triangular muscle covering the upper and posterior part of the neck and shoulders (Figure 5). It is the muscle one feels between the base of the neck and the shoulder and is often involved in cumulative trauma disorders. A number of muscles lie in the neck under the trapezius (Figure 2). The levator scapulae muscle extends from the cervical vertebrae to the superior border of the scapula and dysfunction can result in stiff neck.

Also of considerable importance anteriorly (in the front) in the neck are the three scalenus muscles which can compress the nerves and blood vessels as they pass into the arm. The scalenus anterior muscle attaches to the transverse processes of the third to the sixth cervical vertebrae and to the first rib. The scalenus medius is just posterior to it. The brachial plexus and subclavian artery emerges from the thorax through the thoracic outlet to the arm between scalenus anterior and scalenus medius muscles (Figure 1).

The rotator cuff of the shoulder consists of five muscles which originate at the scapula and attach the greater tubercle of the humerus. The biceps is a long muscle in the front of the arm. It has a short head which arises from the coracoid process of the scapula and a long head arising from the cavity of the humerus. The long head forms a tendinous attachment which is subject to tendonitis (bicipital tendonitis). Distally, the biceps muscle attaches into the radius. The biceps flexes the arm, forearm, and supinates (turns palm up) the hand. The triceps is situated on the back of the arm and is responsible for extending the forearm. It is the antagonist of the biceps muscle (Figure 6).

On the front of the forearm are the flexor muscles of the fingers (Figure 6).[6] There is a superficial group called the flexor digitorum sublimis and a deep group called the flexor digitorum profundus. The sublimis group flexes the second phalanges of each finger while the profundus group flexes the terminal phalanges of each finger. In addition, the flexor carpi radialis and flexor carpi ulnaris are attached to the ulnar and radial sides of the carpal bone respectively and flex the hand at the wrist.

The pronator teres rises from medial side of the humerus and forearm and is responsible for pronating the forearm (turning palm down). It has two heads between which the median nerve passes and is quite important when spasm and hypertrophy can result in entrapment of the median nerve (Figure 6).

On the back of the forearm are the extensor muscles which extend the fingers and the wrist (Figure 7). A deeper group of muscles consist of the supinator which is attached to the lateral epicondyle which is affected in lateral epicondylitis. Three other muscles, the abductor pollicis longus which abducts the thumb, the extensor pollicis brevis which extends it and the extensor pollicis longus are important in that the

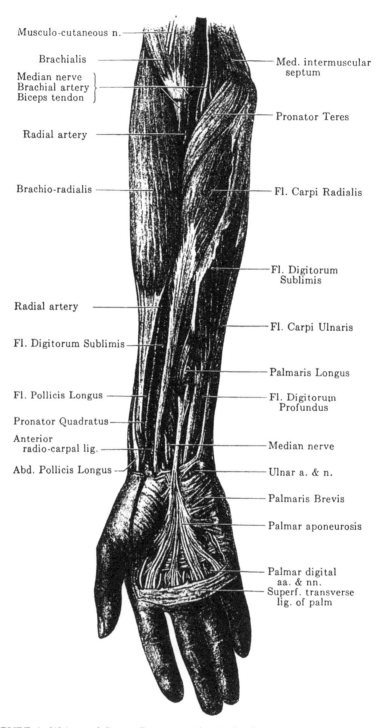

Musculo-cutaneous n.

Brachialis

Median nerve
Brachial artery
Biceps tendon

Radial artery

Brachio-radialis

Radial artery

Fl. Digitorum Sublimis

Fl. Pollicis Longus

Pronator Quadratus

Anterior
radio-carpal lig.

Abd. Pollicis Longus

Med. intermuscular
septum

Pronator Teres

Fl. Carpi Radialis

Fl. Digitorum
Sublimis

Fl. Carpi Ulnaris

Palmaris Longus

Fl. Digitorum
Profundus

Median nerve

Ulnar a. & n.

Palmaris Brevis

Palmar aponeurosis

Palmar digital
aa. & nn.
Superf. transverse
lig. of palm

FIGURE 6. Wrist and finger flexor muscles in the forearm. Grant, J. C. B., *An Atlas of Anatomy*, (Baltimore: Williams & Wilkins, 1947), p. 51. (With permission.)

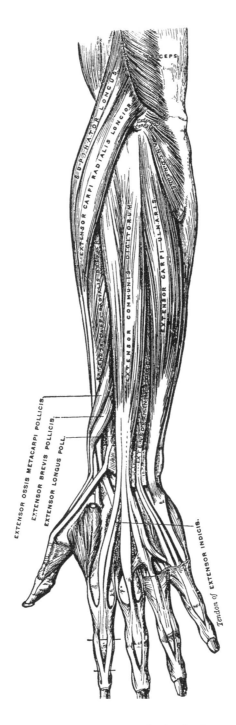

FIGURE 7. Wrist and finger extensor muscles in the forearm. Grant, J. C. B.,
        An Atlas of Anatomy, (Baltimore: Williams & Wilkins, 1947), p. 395.
        (With permission.)

former two end in a common tendon at the base of the thumb and the latter passes just medial to them. The former two are often involved at the base of the thumb in a type of tendonitis called de Quervain's disease (Figure 7).[7]

The above muscles end in tendons which on the palmar side traverse through the carpal tunnel and attach into the fingers. The carpal tunnel is bounded on three sides by the carpal bones of the wrist and on the palmar side by the transverse carpal ligament. Nine tendons pass through the carpal tunnel, in addition to the median nerve (Figure 8). The tendons at the wrist are covered with a common sheath in the carpal tunnel which aids the smooth mobility of the tendons and allows them to traverse closely to one another without interference. This sheath discontinues in some parts of the hand, but then continues as separate sheaths in the fingers (Figure 6).

On the back of the hand, the extensor tendons traverse superficially just beneath the dorsal carpal ligament. A group of several muscles which abduct (move away from the midline), flex, and adduct (move toward the midline) the thumb, comprise the thenar eminence at the base of the thumb. Another group at the base of the little finger called the hypothenar muscles flex, abduct, and adduct the little finger. Lumbricales and interossei muscles are small, deep muscles of the hand, the former rising from the flexor tendons in the palm of the hand. The interossei occupy the intervals between the metacarpal bones and abduct the fingers away from the middle finger while a deeper group of interossei adduct the fingers toward the middle finger.

Nervous System of the Upper Extremity

The upper four spinal cervical nerves unite to form the cervical plexus (Figure 9). The deep branches of the cervical plexus provide innervation to some of the muscles in the shoulders and neck. The lower four cervical nerves combine with the greater part of the first thoracic nerve to form the brachial plexus which goes into the arm. Some branches come off of it above the clavicle, but the major branches unite into three cords which extend into the arm (Figure 8).[8]

Branches from the three cords unite to form the three major nerves of the arm; the radial, median, and ulnar. The median nerve traverses the arm and the forearm and except to innervate the pronator teres muscle, it gives off no branches in the upper arm. Also in the forearm, the median nerve gives off branches which supply the deep muscles in the front of the forearm. Here it also gives off a palmar cutaneous branch which provides sensation to the lower part of the forearm and the palm. This branch does not pass through the carpal tunnel.

In the wrist, the median nerve travels through the carpal tunnel. In the hand, the median nerve provides sensation to the thumb, index, middle, and half of the ring fingers. It innervates to the thenar muscles

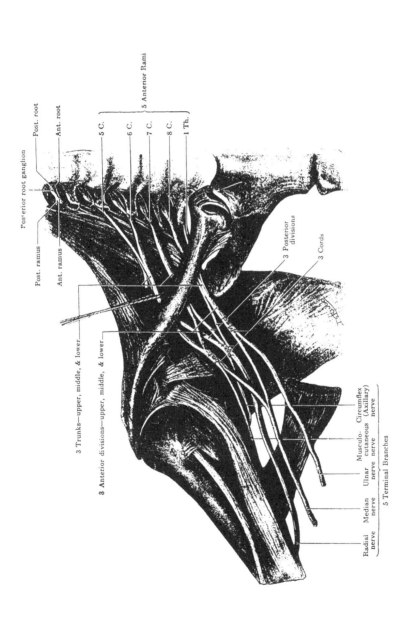

Posterior root ganglion

Post. root

Post. ramus

Ant. root

Ant. ramus

5 C.

6 C.

7 C.

8 C.

1 Th.

5 Anterior Rami

3 Posterior divisions

3 Cords

3 Trunks—upper, middle, & lower

8 Anterior divisions—upper, middle, & lower

Circumflex (Axillary) nerve

Musculo-cutaneous nerve

Ulnar nerve

Median nerve

Radial nerve

5 Terminal Branches

FIGURE 8. The brachial plexus. Grant, J. C. B., *An Atlas of Anatomy*, (Baltimore: Williams & Wilkins, 1947), p. 12. (With permission.)

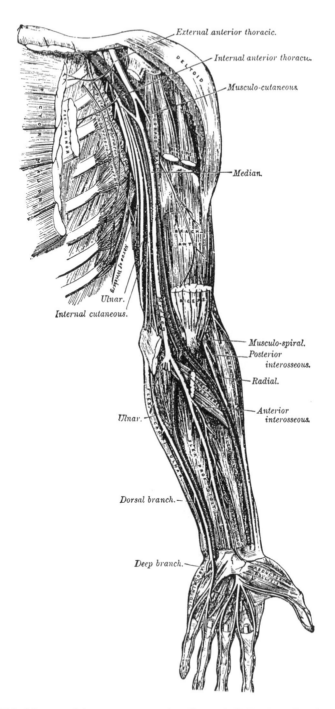

FIGURE 9. Nerves of the upper extremity. Grant, J. C. B., *An Atlas of Anatomy*, (Baltimore: Williams & Wilkins, 1947), p. 767. (With permission.)

and some of the small muscles of the hand (Figure 9).[9] The ulnar nerve runs medially through the upper arm. It is quite superficial at the cubital tunnel just lateral to the medial epicondyle. It passes through the forearm and provides innervation to the deep flexors of the fingers and a number of the muscles in the hand, including the interossei and some of the lumbericals.

The radial nerve is the continuation of the posterior cord of the brachial plexus. It winds around the humerus, passes in front of the lateral epicondyle and divides into a superficial and deep branch. It provides innervation to a number of the extensor muscles in the forearm and to the supinator muscles and to parts of the triceps muscle. It becomes considerably smaller by the time it reaches the wrist.

Circulation of the Upper Extremity

The subclavian artery supplies blood to the upper extremity (Figure 1). It originates on both sides near the sternoclavicular joint. The artery passes up and arches behind the scalenus anterior muscle. It passes between the clavicle and the first rib, then becomes the axillary artery. As it enters into the arm it is called the brachial artery (Figure 1). The brachial artery continues through the upper arm and just below the elbow divides into the radial and ulnar arteries.

The radial artery passes down the radial (lateral) side of the forearm and into the hand, while the ulnar artery passes to the medial side. In the hand, they form a deep and superficial arch called the volar arches, from which the arteries to the fingers originate. The veins collect blood from the hands; a number of veins in the forearm empty into two major veins in the arm. There is a system of superficial and deep veins in the arm which eventually ends in the axillary vein which empties into the subclavian vein.

PHYSIOLOGY OF FATIGUE ON BODY TISSUES

The Conditioning Process

To exercise or work, muscles must contract. In contracting, muscles use oxygen to metabolize carbohydrates, which effects the contraction. Carbohydrates are broken down into carbon dioxide and water. When circulation is adequate to provide the needed oxygen, the muscle is said to be working "aerobically".

When circulation cannot provide oxygen at the rate needed by the exercising muscle, the muscle is in an "anaerobic" condition. Lactic acid then accumulates in the muscle. Muscles that have inadequate oxygenation during work become painful. This immediate pain occurs during the exercise, especially when performing very strenuous work, or work for which the muscles are not conditioned.

Conditioning involves two processes: (1) the cardiopulmonary system, to become more efficient in keeping the blood oxygenated and (2) increase of size and vascularization of muscle fibers.

In addition to the immediate soreness, unaccustomed exercise may cause delayed soreness occurring within 24 to 48 h. It usually goes away in 3 to 5 d. The actual cause of this soreness is not certain. Four hypotheses have been proposed to explain it:[10]

1. Muscle spasm
2. Muscle tears
3. Swelling due to excess metabolites
4. Connective tissue damage

A number of things that occur during strenuous muscle activity may contribute to the delayed soreness. There are two types of muscle contractions: concentric and eccentric. Concentric contractions of the muscle fibers result in shortening of the overall muscle. An example is lifting a weight to the shoulder by flexing the elbow and contracting the biceps. Eccentric contraction occurs when a muscle must support a load during its overall lengthening. An example is the lowering of the load from the shoulder to the hips. No energy is required for the weight to drop. However, to lower it smoothly, muscle fibers must maintain contraction while the overall muscle lengthens.

Eccentric contraction results in more delayed soreness than concentric contraction.[11, 12] Muscle pain also results in spasm causing the feeling of stiffness and tightness that people experience after unaccustomed exercise. One only needs to palpate a painful spastic muscle to demonstrate this effect.

A series of experiments revealed that strenuous work to which the individual was unaccustomed resulted in an increase in the muscle enzyme, serum creatinine kinase (CK) and the muscle protein, "myoglobin".[12] The subjects also complained of soreness. However, when repeating the same task within a 6-week period, the increase in levels of (CK) and myoglobin were less. Soreness was also less, indicating a conditioning process. In other experiments, muscle biopsies were taken after performing strenuous work. Findings included necrosis and myofibrillary disorganization, that is the fibrils of the muscle appeared frayed and disorganized.[13–16]

These acute changes apparently heal completely in a relatively short time. Continuation of the exercise or work makes the muscle better able to withstand the stress.

Another stressor for muscles which can result in immediate pain is that of maintaining the muscle in a static contracted state. This is what happens during isometric exercise. Work above the shoulders can cause prolonged contraction and pain in the muscles of the shoulder and arm.

Muscle soreness and spasm occur in the new worker who is uncon-ditioned for the job. With continuation of the work, this soreness will usually disappear within a few days or weeks, due to the conditioning process. The difficulty in achieving this conditioning will be dependent upon several factors, which include the following.

Rate of Increase in the Activity

Conditioning is much easier to achieve without pain and spasm by gradually increasing activity. It is advantageous for an employee to initially work at slow speed, gradually increasing over several weeks. Increasing of muscle force should also be gradual.

Physical Condition of the Individual

Individuals in good general physical condition usually tolerate new or strenuous work better than those who are not. One of the therapies commonly used in work hardening is aerobic reconditioning. Good physical condition makes an individual more likely to tolerate sustained work even when repetitively using local muscle groups. Good physical condition also results in a sense of well being, and will make an individual mentally more able to tolerate the stress of work.

Muscle strength is another important factor. The lower the percentage of a person's muscle strength required during a task, the less likely the individual is to become injured. Chaffin and co-workers[17–19] have documented this with respect to back injuries. Though strength in relation to required force is probably more applicable to peak loads during heavy lifting, it may also be a factor for repetitive movements. A weak muscle required to exert force repetitively,[1] but well below its capacity, is more likely to fatigue and develop soreness than a strong muscle. The quantitative relationship of strength to the ability to do repetitive tasks requiring force well below capacity is not well studied. For example, no studies are available that show whether grip strength is a predictor of an individual's ability to do repetitive tasks involving the hand.

State of Health

A variety of medical conditions can make it difficult for an individual to achieve the conditioning necessary to perform the job adequately. Many disorders, including the collagen vascular diseases, myasthenia gravis, muscular dystrophy, multiple sclerosis, and weakness from poliomyelitis may potentially affect ability. Arthritic pain and joint swelling may make it difficult for the person to perform repetitive or forceful work.

REFERENCES

1. Gray, H., *Anatomy, Descriptive and Surgical*, (New York: Bounty Books, 1977), p. 528.
2. Gray, H., *Anatomy, Descriptive and Surgical*, (New York: Bounty Books, 1977), p. 338.
3. Gray, H., *Anatomy of the Human Body*, (Philadelphia: Lea & Febiger, 1948), p. 74.
4. Grant, J.C.B., *An Atlas of Anatomy*, (Baltimore: Williams & Wilkins, 1947), p. 2.
5. Chaffin, D. B. and Andersson, G. B. J., *Occupational Biomechanics*, (New York: John Wiley & Sons, 1984), p. 215.
6. Grant, J.C.B., *An Atlas of Anatomy*, (Baltimore: Williams & Wilkins, 1947), p. 51
7. Grant, J.C.B., *An Atlas of Anatomy*, (Baltimore: Williams & Wilkins, 1947), p. 395.
8. Grant, J.C.B., *An Atlas of Anatomy*, (Baltimore: Williams & Wilkins, 1947), p. 12.
9. Grant, J.C.B., *An Atlas of Anatomy*, (Baltimore: Williams & Wilkins, 1947), p. 767.
10. McArdle, W. D., Katch, F. I., and Katch, V. L., "Adaptations with strength training," in *Exercise Physiology*, (Philadelphia: Lea & Febiger, 1986) pp. 385–400.
11. Armstrong, R. B., "Mechanisms of exercise-induced delayed onset muscular soreness: a brief review," *Medicine and Science in Sports and Exercise*, 16:529–538 (1984).
12. Byrnes, W. C., Clarkson, P. M., White, J. S., Hsieh, S. S., Frykman, P. N., and Maughan, R. J., "Delayed onset muscle soreness following repeated bouts of downhill running," *J. Appl. Physiol.*, 59(3):710—715 (1985).
13. Friden, J., Sjostrom, M., and Ekblom, B., "Myofibrillar damage following intense eccentric exercise in man", *Int. J. Sports Med.*, 4:170-176 (1983).
14. Hikida, R. S., Staron, R. S., Hagerman, F. C., Sherman, W. M., and Costill, D. L., "Muscle fiber necrosis associated with human marathon runners," *J. Neurol. Sci.*, 59:185–203 (1983).
15. Warhol, M. J., Siegel, A. J., Evans, W. J., and Silverman, L. M., "Skeletal muscle injury and repair in marathon runners after competition," *Am. J. Pathol.*, 118:331–339 (1985).
16. Kuipers, H., Drukker, J., Frederik, P. M., Geurten, P., and v. Kranenburg, G., "Muscle degeneration after exercise in rats," *Int. J. Sports Med.*, 4:45–51 (1983).
17. Chaffin, D. B., and Park, K. S., "A longitudinal study of low-back pain as associated with occupational weightlifting factors," *Am. Ind. Hyg. Assoc. J.*, 34:513 (1973).
18. Chaffin, D. B., Herrin, G. D., Keyserling, W. M., et al. "Preemployment strength testing," NIOSH Technical Report. Cincinnati, OH, NIOSH Physiology and Ergonomics Branch (1977).
19. Chaffin, D. B., Herrin, G. D., and Keyserling, W. M., "Preemployment strength testing," *J. Occup. Med.*, 20:403–408 (1978).

ADDITIONAL READING

Buckle, P., Ed. "Musculoskeletal Disorders at Work," Proceedings of a Conference Held at the University of Surrey, Guildford 13–15 April 1987. (London: Taylor & Francis, 1987) pp. 118–123.

Gray, H., *Anatomy, Descriptive and Surgical*, (New York: Bounty Books, 1977).

Chapter 2
Workplace Factors

WHAT IS REQUIRED TO OPTIMIZE PHYSICAL FUNCTION?

Before we investigate the causative factors in the development of CTDs, we first need to determine what is required for optimum physical functioning of the soft tissues in the upper extremity. We find that muscles, which provide the movement, strength, and positioning for physical work, require five primary conditions: (1) good circulation, (2) regular movement, (3) leverage, or efficient force exertion, (4) use of the right muscle for the task, and (5) recovery time.

Circulation

Muscle tissue is supplied by a network of blood vessels. These vessels carry blood from the heart to the muscles. Once inside the muscle, the blood vessels become small in diameter, forming a capillary bed where oxygen and fuels are exchanged for carbon dioxide and waste products developed during muscle usage. This exchange is vital for continuing function of the muscle, as a constant supply of oxygen and fuel is needed to maintain efficient and regular use of the muscle tissue. The exchange of carbon dioxide and waste products that develop during muscle use is important: a buildup of these waste products is known to develop muscle soreness and reduced work capacity.[1] Oxygen is the key to repetitive, prolonged use of muscles.

Movement

Movement in the muscle exercises the tissue to maintain strength and flexibility. Muscles that are not allowed to move, such as when a limb is casted, become deconditioned and actually lose strength, flexibility, and size.

Movement also assists in the exchange of fuels, with the waste products given off by the exercising muscle. The system of blood vessels that delivers blood to the muscle tissue is strong and elastic. However, on the return journey away from the muscles to the heart, the vessels, otherwise known as "veins", are not as elastic, and tend to allow blood to pool. Muscular contractions squeeze these vessels and assist the blood in its journey back to the heart. When movement is not allowed, a stationary or "static" muscle allows blood to pool in the large veins. The venous return of blood is reduced, creating a less efficient exchange of fuels in the muscle tissues.

Overall, lack of movement in the muscle impairs circulation and conditioning, and over time will contribute to stiff, sore, and deconditioned muscles.

Leverage for Efficient Force

The effect of a muscle's contraction will depend on its attachment to the bone and joints. When two or more muscles act on a joint to provide movement around that joint, the amount of effort required will depend on three factors: (1) each muscle's location in relationship to the joint, (2) the angle of pull around the joint, and (3) the force developed by the muscle.[2] The study of this action of leverage between muscles, bones, and joints is called *biomechanics*.

In essence, the muscles act as dual pulley systems as they contract and relax to change the line of pull on a joint. Two muscle groups are located around a joint to work in "force couples", one contracting to provide movement in its direction, while the opposing muscle lengthens and relaxes to allow that movement to occur. Muscles must have a certain amount of elongation in order to produce efficient tensile force, which we know as "strength". This position is called the "neutral" position, and allows both muscles in the force couples to deliver the most force with the least amount of energy expended.[3,4] These neutral positions give the joints a mechanical advantage and allow the muscles to perform work with less susceptibility to fatigue.

Therefore, when working repetitively, it makes sense to position the muscles so that they can deliver the force required with the least energy expended, thus minimizing the onset of fatigue. These positions of leverage for the body are illustrated in Figure 1.

Choosing the Right Muscle for the Task

A muscle's ability to produce tension increases with its cross sectional area.[3] Functionally, the smaller muscles in the body are not as strong as the larger, broader muscles. In discussing upper extremity CTDs, we can compare the hand and finger musculature to the muscles of the forearm, upper arm, or shoulder girdle. By cross section alone, the smaller hand muscles are less capable of exerting high forces, and therefore can fatigue rapidly when manipulating objects weighing even a few ounces. The smaller muscles in the body are physiologically designed to perform light tasks, rather than power tasks. Alternately, the larger upper arm and shoulder musculature are capable of repeatedly handling objects weighing several pounds.

To optimize the physical function of a muscle, it must be used in an activity for which it is designed. The hand and fingers are best used in fine precision activities. If all the fingers together are used for a whole hand "mass" grip, then the musculature of the forearm (which both operate the fingers and stabilize the wrist) are active and should be positioned so that they can work efficiently. Heavier larger objects should be manipulated by the upper arm and shoulder musculature and need to be positioned so these muscles can be used.

Recovery Time

The more forceful the exertion by the muscle, the more recovery time it will need in order to repeat that exertion without fatigue or injury. Likewise, if the muscle is placed in an inefficient position, away from the "leverage" positions described in Figure 1, additional recovery time will be necessary.

When a worker uses 100% of a muscle's strength to perform a task, the muscular effort is called a maximum voluntary contraction (MVC). An MVC can be generated by a one time "all that you've got" strength test using a grip, pinch, or other type of force gauge. Job force requirements, as defined by the percent of available strength necessary to lift or move an object, can be classified as "light", "moderate", or "heavy", as depicted in Table 1.[5]

As illustrated in Figure 2, "light" exertion is classified as one that asks an employee to use 30% or less of the available strength (MVC) in that muscle. "Moderate" work requires the employee to use between 31 and 50% of available strength in the working muscle. "Heavy" work requires 80% of available strength. The higher the percent of MVC used, the fewer strength "reserves" remain for the muscle to use when repeating the effort. Therefore, with heavy work, more recovery time will be necessary to ensure that the appropriate exchange of energy supplies and waste products can take place so that the muscle is fueled properly to repeat the heavy effort.

Both forceful and static exertions will require much circulation to the muscles; if not available, the muscles will fatigue rapidly. The required recovery time for heavy work is greater than the actual effort holding time. It takes more time to replace the fuels needed by the muscle for one single heavy exertion than it takes for the effort to occur!

A muscle's MVC varies according to muscle mass and circulation, which is affected by (1) level of previous conditioning (heavy work, weight lifting, etc.), (2) gender, and (3) age. All of these factors relate to the actual physiological cross section of a muscle. The larger this cross section, the more tension a muscle can produce.[6]

Neck - 0 to 30°

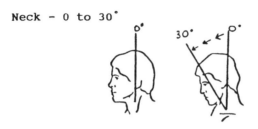

Shoulder - 0 to 45°

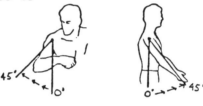

Elbow - 90 to 110°

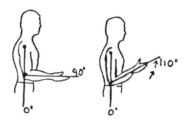

Wrist - 0°

Fingers - Opposing, with fingertips overlapping or slightly apart

FIGURE 1.   Neutral postures, the positions for muscular leverage for upper extremity joints.

TABLE 1. Classification of Muscular Exertion

Effort category	% of capacity (MVC)
Light	0–30%
Moderate	31–50%
Heavy	> 50%

From Rodgers, S. H., "Recovery time needs for repetitive work," *Seminars in Occupational Medicine*, 2(1):19–24 (1987).

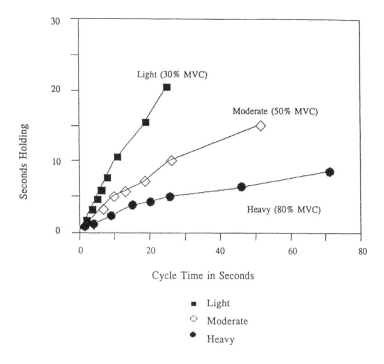

FIGURE 2.   Static work recovery curves. From Rodgers, S. H., Seminars in Occupational Medicine, 2(1):19–24 (1987.) (With permission.)

Previous Conditioning

The strengthening process increases the physiological cross section of a muscle. Therefore, with similar previous activities or strength training, the muscle can produce more tension, thus creating a higher MVC.

Gender

Many studies have shown that females have less muscular strength than males. Laubach (1976) compared nine separate studies of static and dynamic muscle strength measurements of men and women.[6] (See

TABLE 2. Muscular Strength of Females, Compared to Males

Muscular "group"	Average	Range
Upper extremity	55.8%	35–79%
Lower extremity	71.9%	57–86%
Trunk	63.8%	37–70%

Note: ♀ strength, as a percentage of ♂ strength.

From Laubach, L. L., "Comparative muscular strength of men and women: a review of the literature", *Aviation, Space and Environmental Medicine,* (May, 1976), pp. 534–542.

Table 2). Upper extremity strength measurements in women were found to be 35 to 79% of those in men, with an average of 55.8%. Lower extremity strength measurements in women ranged from 57 to 86% of those in men, averaging 71.9%. Trunk strength for women ranged from 37 to 70% of those in men, averaging 63.8%. Dynamic strength tests indicated that women were from 59 to 84% as strong as men, with an average of 68.6%. Overall, in upper extremity strength (grip, forearm, upper arm, and shoulder musculature), the difference between strength capabilities is the greatest.

Age

    As the body ages, a general degenerative process takes place. Typically, muscles lose flexibility and cross sectional size. Generally, older workers will have less strength than younger workers. This is illustrated in studies of adult pinch and grip strength by Mathiowetz et al., as in Tables 3 and 4.[7] Normative data was collected for the grip and pinch strength of female and male adults aged 20 to 75 years (sample of 310 males and 328 females). Their findings showed that the highest grip strength scores occurred in the 25 to 39 years age group. Average pinching grip scores remain relatively stable from 20 to 59 years of age, and gradually decline from 60 to 79 years. This was most obvious on the low scores, with up to a 57% reduction in strength from the youngest age group (20 to 24 years) to the 75-year-old group. In grip strength, a significant difference is also seen between the youngest group (20 to 24 years) and the oldest group (75+).

PHYSICAL STRESSORS AT WORK

    Fatigue, specifically local fatigue, refers to the excessive use of a muscle or body system and is often called "overuse" or "overexertion". Pain in the affected muscle or muscle group is usually a result of fatigue and may be experienced anywhere in the body. A cramp in the hand due to constant use of the fingers is one example of this type of fatigue.

    Fatigue of a specific muscle or muscle group can be classified as either acute (one time incident) or chronic (developing over a long

TABLE 3. Average Performance on Tip Pinch (Pounds) of all Test Subjects

Age	Hand	MEN					WOMEN				
		Mean	SD	SE	Low	High	Mean	SD	SE	Low	High
20–24	R	18.0	3.0	0.57	11	23	11.1	2.1	0.42	8	16
	L	17.0	2.3	0.43	12	33	10.5	1.7	0.34	8	14
25–29	R	18.3	4.4	0.84	10	34	11.9	1.8	0.35	8	16
	L	17.5	5.2	0.99	12	36	11.3	1.8	0.35	9	18
30–34	R	17.6	6.7	0.71	12	25	12.6	3.0	0.58	8	20
	L	17.6	4.8	0.93	10	27	11.7	2.8	0.54	7	17
35–39	R	18.0	3.6	0.73	12	27	11.6	2.5	0.50	8	19
	L	17.7	3.8	0.76	10	24	11.9	2.4	0.47	8	16
40–44	R	17.8	4.0	0.78	11	25	11.5	2.7	0.49	5	15
	L	17.7	3.5	0.68	12	25	11.1	3.0	0.54	6	17
45–49	R	18.7	4.9	0.92	12	30	13.2	3.0	0.60	9	19
	L	17.6	4.1	0.77	12	28	12.1	2.7	0.55	7	18
50–54	R	18.3	4.0	0.80	11	24	12.5	2.2	0.44	9	18
	L	17.8	3.9	0.77	12	26	11.4	2.4	0.49	7	16
55–59	R	16.6	3.3	0.73	11	24	11.7	1.7	0.34	9	16
	L	15.0	3.7	0.81	10	26	10.4	1.4	0.29	8	13
60–64	R	15.8	3.9	0.80	9	22	10.1	2.1	0.43	7	17
	L	15.3	3.7	0.76	9	23	9.9	2.0	0.39	6	15
65–69	R	17.0	4.2	0.81	11	27	10.6	2.0	0.39	7	15
	L	15.4	2.9	0.55	10	21	10.5	2.4	0.45	7	17
70–74	R	13.8	2.6	0.52	11	21	10.1	2.6	0.48	7	15
	L	13.3	2.6	0.51	10	21	9.8	2.3	0.43	6	17
75–+	R	14.0	3.4	0.68	7	21	9.6	2.8	0.54	4	16
	L	13.9	3.7	0.75	8	25	9.3	2.4	0.47	4	13
All	R	17.0	4.1	0.23	7	34	11.3	2.6	0.15	4	20
subjects	L	16.4	4.0	0.23	8	36	10.8	2.4	0.14	4	18

From Mathiowetz, V., Kashman, N., Volland, G., Weber, K., Dowe, M., and Rodgers, S., "Grip and pinch strength: normative data for adults", *Arch. Phys. Med. Rehabil.*, 66:69–72 (1985).

period of time). Acute fatigue is relieved by rest, whereas chronic fatigue occurs when insufficient rest or recovery time exists between work activities. When muscles are not allowed to recover from fatigue, excessive stress is placed on other musculoskeletal components such as the tendons which may lead to tendonitis.

Therefore, the amount of fatigue we experience is directly related to both the *intensity* (amount of force applied) and *duration* (length of time) of the force applied. The intensity of a muscle contraction is more critical in producing fatigue than the duration. Fatigue sets in more rapidly with increased force than with increased duration. Therefore, fatigue may be reduced by decreasing the force load on the muscles (intensity) without changing the duration of work.

TABLE 4. Average Grip Strength Performance (Pounds) of all Test Subjects

		MEN					WOMEN				
Age	Hand	Mean	SD	SE	Low	High	Mean	SD	SE	Low	High
20–24	R	121.0	20.6	3.8	91	167	70.4	14.5	2.8	46	95
	L	104.5	21.8	4.0	71	150	61.0	13.1	2.6	33	88
25–29	R	120.8	23.0	4.4	78	158	74.5	13.9	2.7	48	97
	L	110.5	16.2	3.1	77	139	63.5	12.2	2.4	48	97
30–34	R	121.8	22.4	4.3	70	170	78.7	19.2	3.8	46	137
	L	110.4	21.7	4.2	64	145	68.0	17.7	3.5	36	115
35–39	R	119.7	24.0	4.8	76	176	74.1	10.8	2.2	50	99
	L	112.9	21.7	4.4	73	157	66.3	11.7	2.3	49	91
40–44	R	116.8	20.7	4.1	84	165	70.4	13.5	2.4	38	103
	L	112.8	18.7	3.7	73	157	62.3	13.8	2.5	35	94
45–49	R	109.9	23.0	4.3	65	155	62.2	15.1	3.0	39	100
	L	100.8	22.8	4.3	58	160	56.0	12.7	2.5	37	83
50–54	R	113.6	18.1	3.6	79	151	65.8	11.6	2.3	38	87
	L	101.9	17.0	3.4	70	143	57.3	10.7	2.1	35	76
55–59	R	101.1	26.7	5.8	59	154	57.3	12.5	2.5	33	86
	L	83.2	23.4	5.1	43	128	47.3	11.9	2.4	31	76
60–64	R	89.7	20.4	4.2	51	137	55.1	10.1	2.0	37	77
	L	76.8	20.3	4.1	27	116	45.7	10.1	2.0	29	66
65–69	R	91.1	20.6	4.0	56	131	49.6	9.7	1.8	35	74
	L	76.8	19.8	3.8	43	117	41.0	8.2	1.5	29	63
70–74	R	75.3	21.5	4.2	32	108	49.6	11.7	2.2	33	78
	L	64.8	18.1	3.7	32	93	41.5	10.2	1.9	23	67
75–	R	65.7	21.0	4.2	40	135	42.6	11.0	2.2	25	65
	L	55.0	17.0	3.4	31	119	37.6	8.9	1.7	24	61
All	R	104.3	28.3	1.6	32	176	62.8	17.0	0.96	25	137
subjects	L	93.1	27.6	1.6	27	160	53.9	15.7	0.88	23	115

From Mathiowetz, V., Kashman, N., Volland, G., Weber, K., Dowe, M., and Rodgers, S., "Grip and pinch strength: normative data for adults", *Arch. Phys. Med. Rehabil.*, 66:69–72 (1985).

By understanding the evolution of muscle fatigue and the requirements for optimum physical function, it becomes quite simple to then identify the physical stressors in any given work situation. These physical stressors, also known as "ergonomic stressors" can generally be grouped in three categories: force, posture, and repetition.

Force

Muscles in the body produce the force required to perform work activities. Force is a critical contributing factor to occupational injuries and illnesses. Under optimal conditions, the muscles have the strength capability of producing sufficient force for motion and material handling. However, if the muscle's strength capacity is not adequate for the force requirements of a job, three problematic situations can occur:

1. The force requirement is greater than the muscle's strength, but due to a worker's determination or job requirements, the task is attempted anyway. The result is a muscle strain of immediate onset, the common "overexertion" injury.
2. The force requirement is greater than the muscle's strength, but different from the above example, the task is attempted by "manhandling" and with excessive motion. This extra work (that was *not* intended by the task's designer) requires that additional joints and muscles be recruited, thus imposing additional energy expenditure, an accelerated onset of fatigue, and an inefficient performance.
3. The force requirement is equivalent to a large portion (greater than 50%) of the muscle's strength. The muscle, therefore, is capable of performing the task, but the effort requires so much of its strength capability that little reserve is left for repeated efforts. As the efforts are repeated, less and less reserve is available and the muscle becomes fatigued.

The method of grasping and moving objects is critical to the impact of *force* on the development of upper extremity fatigue and subsequent CTDs. It has been shown that the power grip (whole-hand fisted grip) is five times more powerful than a pinch grip.[8] By using a larger muscle mass (five-finger power grasp) to grasp an object, rather than smaller muscles (two-fingertip pinch), it is possible to reduce the strength requirements of a task. In a power grasp, the strength needed to move the object is distributed to all five fingers, and does not require "manhandling" or excessive movement. Instead, each muscle operating the five fingers is working well under capacity, contributing to the total force required of the task. Conversely, in the pinch grip, the object weight is concentrated onto two smaller muscles and requires a larger percentage of those muscles' strength capacity.

Ćatović et al. (1989) found this to be true when evaluating the design of dental tool handles and the grasp required in their use.[8] This group also found that hand pinch grip strengths were higher when the arm's weight was externally supported. This suggests that jobs involving the recruited muscular effort to maintain the position of the arm during the performance of hand intensive tasks require more energy and allow less grip strength, which could result in an inefficient, fatigue-prone performance.

Many times, external forces can produce "wear and tear" on the body. These are classified as "impact" stressors as the force acts on a concentrated area of the body, disrupting the circulation and creating a "bruising" or contusion type of injury. Examples include

- Using tools with handles that are too small, short, or have squared, sharp edges, creating pressure points on the hand and fingers. For example, forcefully opening scissors can result in swollen, bruised fingers. This is common in the apparel, electronics, and poultry industry.

- Hitting, pushing, or pulling objects with the "butt" or palm of the hand.
- Holding vibrating objects and tools.
- Resting the lower arm, elbow, or wrist on a table or machine edge.

Posture

If the body does not assume biomechanically efficient positions for the joints and muscles, "extreme postures" result. Extreme postures negatively impact the muscular leverage and body positioning necessary for the appropriate muscle to be used in a specific task. When extreme postures are assumed, the available strength to that body part decreases and the onset of muscle fatigue is accelerated.

Wiker (1986) investigated the effects of relative hand movement and fatigue.[9] He found that movement performance was reduced by up to 30% when hands were positioned above shoulder level, holding loads comparable to single hand tools, with long work-to-rest ratios in a given work cycle. The reduction in movement performance associated with these overhead postures was significant: the subject's maximum performance capabilities *failed* to meet the predicted performance using a Predetermined Time System (MTM-1). From these and other findings, Wiker concluded that limitations in movement performance when working overhead were largely due to fatigue in the shoulder musculature. Another study (Jonsson and Hagberg, 1974) found that working heights greater than 75 in. imposed force loads in the neck and shoulder three times greater than those created at work heights of 43 to 47 in.[10] The greater height required the neck and shoulder to move away from neutral positions.

Many studies have defined, by joint angle, those postures that are "extreme" and impose stress to the muscles, as documented by reduced work efficiencies and onset of pain and fatigue.[11–16]

Simply stated, extreme postures are those that are not efficient, neutral postures. A summary of extreme postures, as defined in previous studies, is included in Table 5. Suzanne Rodgers' work on hand position and its effect on grip strength is an excellent example of the effect of posture on the muscle's ability to do work efficiently.[5]

As shown in Table 6, we see that extreme wrist flexion (60° from neutral) can result in a 55% reduction in grip strength, leaving an extreme deficit of strength resources. Fatigue will occur more rapidly than if the hand and finger muscles had been used while placing the wrist in a straight position.

Extreme posture can also initiate a "chain reaction" from the toes up to the neck and fingertips. If the lower half of the body is improperly positioned, the upper half of the body must assume extreme positions to "balance" and compensate for the poor position of the legs, hips, or

TABLE 5. Summary of Extreme Postures

Body joint	Extreme positions resulting in fatigue
Ankle	• Extreme dorsiflexion (toes up) • Extreme plantar flexion (pointed toes)
Knees	• Locked, straight • Flexed, with angle less than 90° (between ankle and thigh)
Hips (seated)	• Greater than 110° or • Less than 80°
Back	• Flexed, side-bent, or extended greater than 20°
Neck	• Rotated • Bent to side, back or forward greater than 20°
Shoulders	• Elevated (shrugged) • Flexed or abducted, with elbows greater than 150° away from the body • Arms extended behind the body
Elbows	• Flexed, with angle less than 80° between upper arm and forearm • Extended, with angle greater than 120° between upper arm and forearm
Forearms	• Rotated with palm up/down, during forceful efforts
Wrists	• Ulnar deviation greater than 45° from neutral • Radial deviation • Flexion greater than 30° • Extension greater than 15° from neutral
Fingers	• Assuming wide grasp (fingertips greater than 1 in. apart) • Pinching

spine (Figures 3 and 4). Seated posture is especially (Figure 5) susceptible. The following description of a typical scenario illustrates the chain reaction effect:

> A short female worker seated in her work chair is unable to touch the floor without pointing her toes or wearing high heels. To compensate, the worker slips forward in the chair, allowing her feet to touch the floor and provide a sense of balanced weight distribution. When slipping forward in the chair, her spine becomes rounded as she leans backward against the back rest, attempting to also get some support from it. There is inadequate clearance for her knees, so she sits on one hip to move the knees to the left and out of the way. This redistributes more weight on the left hip, creating yet another unbalanced situation for the spine. The spine must curve back to the right to allow the worker's upper body to remain squared to the work she is performing. This causes the right shoulder to drop and the left shoulder and elbow to rise. The neck, then, must also tilt to the left to balance the upper body.

TABLE 6. Factors Affecting Grip Strength, Based on Data Collected in Student Laboratories and Industrial Training Courses By S. Rogers 1982–1985

Condition	Percentage of power grip
Power grip—2 in. span	100
with 25° radial deviation	80
with 45° extension	75
with 45° ulnar deviation	75
with 45° flexion	60
with 60° flexion	45
with rubber gloves	80
with cotton gloves	75
with heavy leather gloves	65
Power grip—3 in. span	70
—1.5 in. span	75
—1.25 in. span	55
—1 in. span	40
Pinch grip	15–25

From Rodgers, S. H., "Recovery time needs for repetitive work," *Seminars in Occupational Medicine*, 2(1):19–24 (1987).

FIGURE 3. Balanced seating: both sides of the pelvis are supported with resulting balance in the upper body.

The overall result of this scenario is an "S" posture from the hips to the neck with one shoulder raised. The muscles are no longer biomechanically balanced. Low back, midback, neck, and shoulder fatigue can result, followed by complaints of headaches and back pain. Two

FIGURE 4.   Unbalanced seating: one hip is elevated creating compensatory extreme postures in the spine and shoulders.

FIGURE 5.   Illustration of "chain reaction" compensatory postures.

seemingly unrelated issues create this postural chain reaction: no support for the feet and inadequate knee clearance. So, in this scenario, an extreme posture in the feet could create a headache at the end of the workshift.

Repetition

Repetition, unfortunately, has earned an inappropriate reputation for being *the* causative factor in CTDs. The early name for CTDs, "repetitive motion injuries", indicted repetition as the single most influential stressor in their development. This is largely based on a study performed at the University of Michigan.[17] This study generically determined that a cycle time of 30 s or less, or a concentration of greater than 50% of cycle time spent performing the same task (fundamental cycle), as the determining factors for defining a job as "highly repetitive". However, the study also noted that force was a major factor in the development of CTDs, with the *combination* of force and repetition having the strongest predictive value of the onset of CTDs. This finding has more practical significance than simply indicting repetition alone as the primary risk factor for CTDs.

After exertion, muscles require recovery time. As described earlier, heavier exertions require more recovery time. If adequate recovery is not allowed within the total cycle time of a repetitive job, then this lack of rest will result in a state of cumulative fatigue at the end of the workshift. Thus, the repetition becomes the secondary stressor. This situation can be resolved in two ways: (1) reduce the required force so that the allowed recovery time is adequate for the exertion, without changing the cycle time or number of repetitions, or (2) maintain the force requirement and allow more recovery time by reducing the repetitions, increasing the cycle time, or adding more labor. Obviously, in competitive world markets, it becomes increasingly difficult for management to "slow down" the production process or add direct labor costs to the product. Therefore, option 1, minimizing the force requirement of the job, can be just as effective while more palatable to the economic pressures of today's market.

A closely related stressor, *work concentration*, is often confused with repetition. A job that is simply repetitive, requiring that a job be performed repeatedly in short cycles, does not necessarily place the worker at risk of developing a CTD. Rather, the concentration of the work, or lack of task variety, may contribute to local muscle fatigue. A concentrated job consists of few work elements, requiring a single muscle to perform a singular movement in the same direction all day everyday. The lack of task variety, as dictated by the job's task design, results in "overuse" of this muscle, thus eliciting fatigue. Jobs are often designed in this manner in the name of "efficiency".

Static loading is a commonly used term to describe an unchanging body position of prolonged duration. Typically, it involves all three stressors: (1) an *extreme posture* created by the worker's positioning around a workstation that does not "fit", (2) *force* associated with prolonged holding of the limb's weight and the weight of the object manipulated, and (3) the absence of *recovery time*.

The body functions best when movement is allowed in the muscles and joints. Discomfort is often experienced due to fatigue from holding muscles contracted in a fixed or awkward position for extended periods. This fatigue is a result of inadequate circulation to the muscle, as the holding tension of the muscle "chokes" out the blood supply.

Work situations involving the following duration of force are considered static postures:[18]

- High effort   8 lb load held for 10 s
- Moderate effort   4 lb load held for 60 s
- Light effort   <2 lb load held for 4 min

Examples of static work include holding the arms up, elbows out, bending the neck to the front or side, holding a tool in the hand for prolonged durations, and any other fixed position that is held for an extended period in a fixed position which hinders adequate blood supply. In contrast, work that allows muscles to contract and relax enhances circulation, which helps to avoid, or at least decrease, the rate of fatigue onset. Static loading commonly occurs in the neck and shoulder when a worker is positioned too low relative to the worksurface, resulting in raised arms and static loading in the shoulder musculature, or too high relative to the worksurface, requiring one to look down at the work, creating a static load in the neck musculature.

The Root Cause

Any feature of the work environment or the job design can negatively impact the worker's physical ability to perform the job efficiently, creating undue fatigue and ultimately a CTD. This section describes the more common offenders.

Kroemer (1989) grouped workplace offenders by citing seven "avoid" scenarios.[19]

1. Job activities with many repetitions
2. Work that requires prolonged or repetitive exertions of more than 30% of the operator's muscle strength available for that activity
3. Placing body segments in an extreme position, such as severely bending the wrist
4. Work that makes a person maintain the same body posture for long periods of time
5. Work in which a tool vibrates the body or part of the body
6. Exposure of working body segments to cold, including air flow from pneumatic tools
7. Combinations of the conditions just described

These seven scenarios depict the basic stressors of force, posture, and inadequate recovery time. However, it is a rare occurrence that a stressor acts alone to create a CTD. In reality, the "root cause" may be related to a combination of many workplace/job design features.

FIGURE 6.   Seating: short legs with inadequate support from seat back or floor.

SOURCES OF ERGONOMIC STRESS

Workplace Layout and Design
Seating
Lack of Adjustability. All workers are not of the same shape or dimension. Two people with a stature of 70 in. may have two completely different leg and torso lengths, resulting in one's ability to reach the floor and the other's ability to reach a high shelf. The lack of seating adjustability in its height and components (back rest, arm rests, footrests, etc.) will force the worker to assume extreme positions in an attempt to adjust the body to the chair and worksurface.

Seat Pan. If the seat pan is too long, workers with short legs will not be able to utilize the back rest and keep the feet flat on the floor, as illustrated in Figure 6. The short worker typically slips forward in the chair in an attempt to place the feet flat on the floor, resulting in inadequate back support. A seat pan that is too long can also create a mechanical impact to the back of the knee, impairing circulation to the lower leg (Figure 7). If a seat pan is too short, the worker's legs are exposed to mechanical impact at the back of the thighs, which can impair circulation, and create pressure on the major nerves that supply the lower extremity (Figure 8).

FIGURE 7.   Seating: short legs with long seat pan resulting in mechanical impact to the leg.

FIGURE 8.   Seating: long legs with short seat pan resulting in mechanical impact to the leg.

FIGURE 9.   Seating: back rest out of position providing inadequate support for the lumbar spine.

Back Rest. If the back rest does not fit against the lumbar curve of the back, and is not adjustable to the various torso heights of the working population, then the back will not be adequately supported (Figure 9). Extreme postures and muscular fatigue result from the lack of external support for the weight of the body.

Padding. If a seat is not padded, the torso (pelvis and spine) must absorb the impact of the body's weight. If a seat is padded, the impact of the body weight can be distributed, which results in less fatigue of the pelvic and back musculature. Additionally, rounded padded "waterfall" edges on the seat minimize the chance of pressure points at the back of the knee or thigh.

Footrests. Footrests are directly related to chair height adjustability. If a chair's height is adjustable, then footrests should be available for those workers who must adjust the chair height upward in order to comfortably reach the worksurface. If no footrests are available, the legs will dangle, resulting in increased pressure at the back of the thigh and knees, imbalanced weight distribution and increased force requirements on the torso (Figure 10).

Arm Rests. Arm rests can become a hindrance if there is insufficient clearance under the worksurface for both the chair and the arm rests.

FIGURE 10. Seating: inadequate foot support results in imbalanced weight
distribution and increased force requirements.

This results in positioning away from the workstation with extreme
shoulder abduction and forward trunk flexion postures to reach the
work (Figure 11). Arm rests can be a welcomed feature if the arms
require support while the hands perform a precision task. If the upper
arm is not to be used, providing a prop will reduce fatigue in the
supportive shoulder, neck, and upper back musculature.

Type of Seating. Not all chairs are correct for all situations. An incorrect
seating option can become more of a hindrance and stressor than an
assistive device. The following principles should apply when choosing
the type of seating to use in an operation:

- **Standing** — If the job requires reaching, lifting, pushing, pulling,
  carrying, or working from one worksurface to another, then it is most
  appropriate for the worker to stand. This allows more mobility in the
  body and provides the worker the opportunity to assume neutral joint
  postures when performing these varied tasks.

- **Traditional chair with backrest and seat** — The traditional chair or
  stool should be used for jobs that are fine and precise in nature,
  requiring the eyes to be focused on the work and the hands to be
  placed in front of the body while manipulating the equipment or
  materials. The energy expended to hold the body in position is
  minimized, thus reducing overall body fatigue. If the worker is expected

FIGURE 11. Seating: armrests can be a hindrance creating extreme postures
in the shoulder and trunk.

to reach in and out of bins placed to the side or behind the chair, or
on an overhead shelf beyond the worker's reach, extreme shoulder
and back postures can result. The chair then becomes a hazard to the
operation, regardless of its "ergonomic features".

- **Sit/lean stools** — When seated or standing at a workstation, a design
  conflict may occur between the need for visual acuity and hand/arm
  strength. Sit/lean stools are hybrid seating options that allow the
  worker to assume a half-standing and half-seated position. This type
  of seating is most appropriate when the worker must do a combination
  of tasks: fine precise work with an occasional reach or material handling
  activity. This allows the body to be supported during the fine activity
  and allows a half-standing position for ready movement onto the feet
  when necessary.

Worksurface

Height. If the worksurface is too low, then the worker must assume
extreme forward trunk and neck flexion positions to reach the work.
Alternately, if the worksurface is too high, then extreme shoulder
abduction, and possibly low back extension, postures result. Extreme
postures associated with work height are compounded by the fact that
they are typically static.

Obstacles. If a ledge, fixture, bin, or other obstacle is located between
the worker and the work, then the worker must assume extreme reach

postures affecting the trunk, shoulder, and elbow, to pick up, process, or aside the work around these obstacles. This ultimately adds unnecessary carrying distance and handling time, robbing valuable seconds of recovery time within a work cycle.

Orientation. If the worksurface is horizontal and the worker is applying a process to an object placed horizontally on the flat surface, the neck and wrist must deviate downward, resulting in extreme postures in these joints. Many work surfaces are angled to improve the visual angle and reduce these extreme postures on the wrist and neck.

Dimensions. The dimensions of a workstation's layout affect critical reaches. Reach distances relate to extreme postures in the trunk, shoulder, and elbow, and also contribute to unnecessary carrying distances. The result is a combination of extreme posture and inefficiently used cycle time. This can especially affect workers with short torsos and arms. For frequent tasks, reaches are considered extreme if they exceed 14 to 18 in. from the shoulder. Reaches can be as great as 24 to 36 in. away from the body, but these are for infrequent tasks, and are generally not efficient.

Workspace
Leg Room. If inadequate leg room allows no space to place the feet, knees, or thighs, extreme postures from the hip to the fingertips can result. This is generally seen when obstacles (motors or supply boxes) exist under tables, or incompatibility between worksurface height adjustment and seating adjustability leaves no thigh clearance space.

Flooring/Footrests. If workers must stand continually in one place on cement or other hard surfaces, the static loading on the lower extremities can fatigue the feet, knees, hips, and low back. Footrests allow weight distribution when standing with one foot "propped" up. The absence of antifatigue matting and footrests contribute to static standing fatigue.

Material Handling. If surfaces from which materials are transferred from one to another are not level, unnecessary lift distances add carrying time and weight to the task. Material handling that requires lifting, carrying, and traveling on foot results in prolonged static holding with the upper extremity, and minimizes the amount of recovery time available.

Machine
Displays and Controls. The location of machine displays can result in extreme postures in the neck and shoulders if not properly placed in clear view at eye level (see Figure 12). Likewise, controls should be

FIGURE 12. Office equipment with "stacked" VDT components creating extreme neck postures.

positioned so that the hands can depress the button between elbow and hip heights allowing neutral shoulder and arm postures. For example, VDT/office equipment is often stacked on any worksurface without regard to screen or keyboard height.

A general concern is the overall fatigue resulting in the use of a machine and its effect on safety hazards. Fatigued hands and body will result in clumsy motions and may create unsafe acts when working around moving parts. Machine guarding may create an obstacle and create extreme arm positions.

The effort and position required to operate foot or hand controls can impose the combined stress of force and extreme posture on the finger(s) and ankle. Generally, foot pedals should be avoided in standing operations as they create unbalanced weight distribution and poor postures in the lower extremity and torso.

Speed. Machines and conveyors cycle at a speed that is either controlled by the worker or externally paced. Generally, machines with short cycle times or conveyors that feed discrete short-cycle assembly operations may impose insufficient recovery time on the worker, resulting in increased fatigue and an inability to "keep up". For a more detailed description of this stressor, consult discussion under The Job Design, Work Pace and Scheduling.

Tooling[20,21]

Effort Required to Operate. Tools are utilized when the operation calls for greater force and/or speed than a manual effort can provide. Tools provide this by allowing an interface between the hand and the working surface for leverage, or by power (air, electricity, hydraulics). The effort required to operate a tool is directly related to the efficiency of the tool's action, the reaction of the tool onto the body, and the weight of the tool.

A tool's efficiency depends upon proper maintenance. Tools that are not maintained on a regular basis become jerky and inconsistent, requiring more effort from the operator. The working head of a tool must be maintained to perform its action consistently. The best example of the relationship of proper maintenance and ergonomic stress is reflected in the use of knives and scissors in the red meat and poultry industries. The greatest stressor associated with these tools is related to the sharpness of the blade and the ability of the blade to maintain a sharp edge. If sharpened daily and regularly "steeled", knife and scissor blades stay sharp and require minimal force in their use. Dull knives and scissors result in hacking, sawing, and increased repetitions to perform the cuts required, adding both force and repetition stressors to the activity.

Reaction forces on the body require that muscles be used to stabilize the joint as the tool "kicks back". This requires not only strength to handle the tool's weight and to operate its controls, but to also "set" against the reaction forces. This type of force is seen in torque reaction, impact guns, and vibrating tools.

Tool weight imposes effort requirements for their operation. Tools weighing more than ten pounds should be handled by two hands, or a counterbalance system, as this weight is too great for a single arm and hand to lift and operate. Frequently handled tools impose high force requirements if their weight exceeds two pounds. Power hose weight also contributes to the force requirements. Tools that are used continually, resulting in a constant grasping of the handle, impose a static loading stressor on the hand and fingers, especially if the tool is not counterbalanced. Tools with their center of gravity located greater than 10 in. in front of the handle will also impose inefficient leverages and increase strength requirements to physically counterbalance them.

Posture Required to Operate. A wide variety of tools are available with just as many different tool handle configurations. In-line, pistol grip, two-handled, and palm grasps are all common handle designs. A fallacy exists that "pistol grip" handles prevent the development of CTDs, specifically carpal tunnel syndrome. The handle orientation is less important than the posture required to operate it. This posture will always reflect the orientation of the surface upon which the tool is

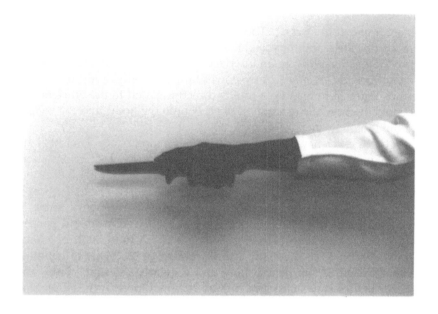

FIGURE 13. Precision grip.

working, and the design of the tool's working head and handle relationship. Pistol grip orientations are appropriate for working surfaces
that are (1) directly vertical to the operator and above hip level, or (2)
horizontal and below hip height. If, however, the worksurface is horizontal and oriented at or about elbow height, the same pistol grip will
result in extreme ulnar deviation. Therefore, if the tool and worksurface
orientation is not considered initially, the best designed tool handle can
still result in deviated wrist postures. This results in weakened grip
strength, which will increase fatigue with repeated handling of the tool
and its weight.

Additionally, the grip posture is dependent on the level of precision
required of the task. Tools that are used in precision tasks require that
the working shaft be aligned with the first three fingers of the hand:
thumb, forefinger, and third finger. This adds fine precision control to
the tool as it is being operated. Power activities such as sawing, grinding,
buffing, and torquing require a five-finger "mass" grip. Therefore, two
types of grips can be identified

- **Precision grip** — This grip requires the fourth and fifth fingers as an
  "anchor", with the thumb assisting; and the second and third fingers
  act as a precision "guide", in line with the working head or shaft of
  the tool (Figures 13 and 14).

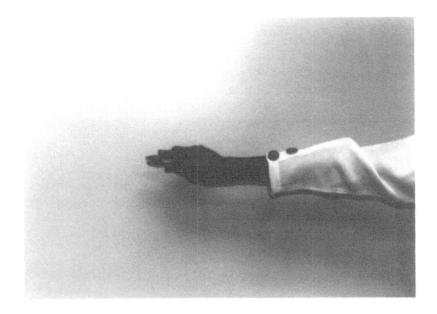

FIGURE 14. Precision grip.

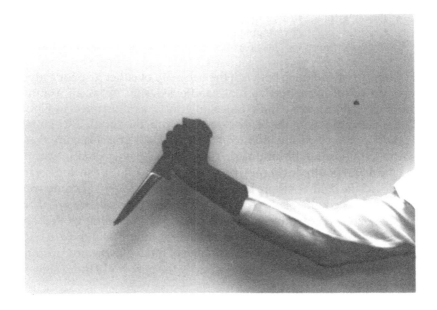

FIGURE 15. Power grip.

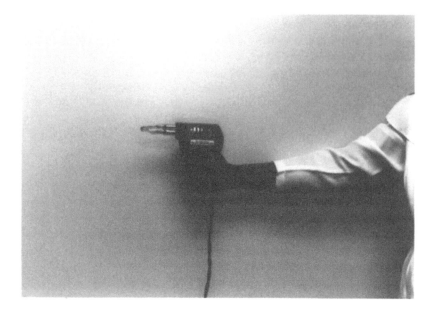

FIGURE 16. Power grip.

- **Power grip** — All five fingers are flexed and closed together, with fingertips slightly overlapping with the thumb, which acts as an anchor (Figures 15 and 16). This provides maximum power as the grip is held.

Failing to distinguish between the necessity of either grip for a given task can result in ineffective tool handle redesign. A recent example involves that of knife handle redesign efforts, as illustrated in Figure 17. Knife handles with the blade extending from the bottom ("U-knife") or the top of a vertically oriented handle ("pistol-grip") require that a power grip position be assumed. These knives were tested in various red meat and poultry deboning processes which require the knife blade to "clean" the bone contours for maximum meat yields. The contouring action with the knife blade required a high level of skill and precision. The new knife handle designs failed miserably in gaining worker and management acceptance. The handle-blade configuration thwarted the worker's ability to assume a precision grip, forcing awkward postures and resulting in inefficient yields — exactly the opposite of the designer's intentions!

Handles. Handles that are too wide or narrow will result in inefficient use of finger and hand musculature and result in increased fatigue.

FIGURE 17. Wide handles do not allow a whole-hand grip. Handles are forced out onto the fingertips.

FIGURE 18. Narrow handles require overflexion of fingers.

FIGURE 19. Short handles can create pressure points in the palms.

Handles that exceed 1.5 in. in diameter or a 3-in. span between two handles (plier and other two-handled tools) will result in wide fingertip grasps, and effectively require a pinch grip by the fingertips (See Figure 17). Alternately, two-handled spans less than 2 in., or diameters less than 1.5 in. result in overflexed fingers, creating palmar cramping and the potential for "trigger finger" tenosynovitis (See Figure 18).

Handles with less than 4 to 5 in. in length can also impose pressure points in larger hands (Figure 19).

Handles that are metallic without nonslip rubber, leather, or synthetic coverings will reduce the frictional resistance and texture, imposing greater grip strength requirements.

Controls. Power tools require that finger controls be utilized. One finger trigger, thumb, or fifth finger controls can impose overuse on those single fingers if used repetitively. Distributing the force requirements of tool control to at least two fingers will reduce overuse and distribute the force requirements.

The Job Design

Methods and Training
Worker Technique/Methods. Industrial engineers are usually responsible for identifying "correct" motions and methods, based on efficiency standards. However, many workers are allowed to improvise, given

their ability to get the job done within required time and quality specifications. Unchecked methods often prove to be *the* primary root cause of a CTD, as excessive motions, improper positioning, unnecessary extreme postures, and poor maintenance habits all create ergonomic stressors in a job.

Improper methods usually result from unstructured training efforts for newly hired employees or those cross training or transferring to new departments. The "quickest" workers are usually recruited to act as on-the-job trainers, with little regard to whether these workers reflect the most efficient and ergonomically correct techniques. Bad habits are picked up from the first day of training and are difficult to break, once learned. Common method "mistakes" from an ergonomics perspective are reaching, exteme and repetitive motion, and concentration of force.

- **Reaching** — Workers may increase reach distances by failing to adequately locate the correct equipment and set up the workstation before starting work. This is often associated with tight production schedules, inadequate maintenance departments, inadequate supplies, or loose administrative efforts to enforce worker punctuality for set-up prior to beginning the operation.

- **Extreme and repetitive motion** — Excessive motions are characteristic of a "working harder than one has to" or "just making sure" mentality. If inadequate training or extremely tight quality standards exist, workers will physically repeat a motion to ensure quality standards are achieved. Also, extreme motion exists when an employee rushes through a job to make or exceed production standards, as in piece-rate compensation. The worker perceives that increased activity equals increased production, and this sets the entire body into motion with rocking and flailing arms. This is typically seen in sewing machine operators in the apparel industry.
  Tight quality standards often result in the operator repeating motions to ensure that the quality standard has been met. A good example of increased motions and yield expectations is seen in the poultry industry. A standard task procedure for thigh deboning (removing meat from the thigh bone, with a yield goal of no meat remaining on the bone) may state that only five cutting motions are necessary to adequately remove the meat from the bird's thigh bone. However, on a production line, the workers performing thigh deboning tasks may be making eight or more cuts to achieve yield expectations. This results in a 50% increase of motions in a given cycle time, robbing from the scheduled recovery time established in the line speed.

- **Concentration of force** — Employees may be overusing a single finger rather than employing the use of several fingers in a grasp or stabilizing motion. Fingers may be used in preference to small tools, to eliminate the need to pick up and handle the tool. The edge of an object may be grasped and moved with a pinch grasp vs a whole-hand grasp, in an effort to "save time". All of these activities impose increased force requirements and a concentration of those force requirements on single small muscle groups, resulting in overuse and fatigue.

Productivity Expectations. Overuse often begins imposing excessive fatigue on new workers when not allowing them time to adequately condition to the activities. Too often, sales dictate short lead-time production levels, which result in "warm body" hiring techniques, placing inexperienced and unconditioned employees in the jobs, who then are expected to work at full productivity. Unable to learn skill and speed simultaneously, workers become frustrated with their attempts to perform well, resulting in the wasteful methods previously discussed.

Training Program
Standard Procedures. Refusal to invest the time and energy in setting up procedures, standard methods, and documentation results in on-line trainers "winging it" and leads to inconsistency and inefficiency in new hires.

Trainers. Refusal to give the lead trainer and on-line trainers the appropriate status results in a "why should I listen to him/her?" attitude. Status provides another niche in the hierarchy for hourly employees, and promotes a desire in hourly employees to exhibit good technique and cross training for promotion purposes.

Refusal to hire a training supervisor and requiring that on-line trainers report to a production supervisor results in the trainers giving production top priority instead of skill acquisition. Also, this hierarchy typically degenerates to the utilization of trainers as utility or fill-in employees.

Discipline Issues. Many employers lack reward systems for poper job performance or disciplinary procedures for dealing with employees who continue to practice nonstandard methods and resultingly become safety and ergonomic "repeat offenders". This results in inconsistency and reduces the impact of the entire training effort, and hurts both the company and the employee.

Inappropriate Use of Light Duty. Training workers in an unrelated "light duty job", utilizing different motions and procedures from the job for which the worker is being trained, does nothing to condition that employee for the job. Supervisors and production managers become frustrated when the worker fails to perform adequately. Inefficient methods and overuse in the musculature often ensues once the employee is placed on the actual job.

Task and Activity Mix

If the job is designed so that a single task is performed over 50% of the work cycle, then it is considered "concentrated" and leads to the potential for overuse of a single body part, joint, or muscle group. Occasional tasks that break up the rhythm and add variety to the overall work cycle assist in offsetting the concentration of work.

A good example of this is reflected in the comparison of the early typist versus today's "word processor". When a typist processed documents and work materials, every line required a carriage return, interrupting the keystroking activity. Every page had to be put into the carriage by hand, further interrupting the keystroke activity. In present day word processing, a worker can simply stare at a screen or document while the fingers endlessly perform keystroke motions. Even corrections can be made with a single keystroke, as well as carriage return, scrolling pages up or down, etc. Therefore, by automating the office, the work has become more intensive and concentrated, focusing all of the work onto the fingertips and hands, and increasing the potential for overuse and fatigue.

Overtime

Overtime is generally used to offset seasonal increases in production demand. Workers perceive work week overtime as "positive" as it will increase pay "temporarily" and avoids cutting into weekend leisure time. Most people would prefer extending a work day versus extending a work week into the weekend. However when working over an 8-h day, fatigue can set in, especially moderate to heavy physical or mental work. Over many weeks, a loss of personal time is perceived by the workers.

Continued overtime can result in decreased hourly production.[22] If adding 2 to 4 h per shift, production has been shown to fall 5 to 10% *below* the inspected increase of 25 to 50%.[22] This can be directly attributed to fatigue. Overtime has been shown to increase the cost per unit produced. Overtime has also been associated with worker dissatisfaction and increased incidence of CTDs.

To offset the fatigue, energy requirements must be decreased by adding breaks or reducing the speed of the process. However, this is rarely considered when the decision is made to work overtime. In self-paced operations, the production per hour may also decrease.

Hours per Week

CTDs may become problematic for continuous processes that compress a 40-h work week into 3 or 4 work days. The 12-h shift performed 3 d in a row may be susceptible to the same fatigue concerns as with overtime scheduling. This is directly related to the recovery time needs that should occur *within* a given work cycle, period, or day. Waiting for recovery time to occur after three consecutive 12-h shifts may be inadequate to offset the accumulation of fatigue that could result in myalgia and tendonitis. Longer "weekends" of 3 or 4 d may increase the potential for moonlighting. Instead of the worker utilizing the nonworking days for recovery and leisure, the worker robs himself of the much needed rest by working two jobs. Additionally, an injury or illness caused by the second job may be aggravated by the work performed for the primary employer, creating workers' compensation liability issues.

Standards/Incentive Pay

Incentive pay or "piece rate" can contribute to CTDs by reducing recovery time availability. These operations are usually self-paced with the operator receiving additional pay for working above standard production expectations. If standard production expectations were established at 100%, factoring in recovery time needs, then fatigue may develop in an employee that produces more than 100%.

Additionally, pacing becomes a key issue. For example, if an operator begins at a 130% productivity rate at 7:00 a.m. and ends the day at a 90% rate, she may have only averaged 112% over the entire shift, which forms the basis for her earnings. However, if she notices that a particular body part becomes extremely fatigued or sore at the middle of the shift, she may be suffering from improper pacing and inadequate recovery time; she earns only 112%, in spite of her initial but sporadic 130% effort. Would it not make more sense to instruct this employee in proper self-pacing, to earn maximum income while minimizing the potential for midday fatigue? If she were to begin at or around 112%, she may be able to generate the same earnings with a steadier and more relaxed pace, allowing more recovery time *throughout* the day. This, hopefully, will allow her to experience *less* fatigue *later* in the day, and may offset a long-term disability that would accumulate over the course of several weeks, months, or years at the previous break-neck pace.

Machine Pacing

Work can either be self-paced or machine-paced (external). Externally paced work creates time and performance stressors that do not exist when a worker has control over the pace. In order to "stay ahead" and reduce the chances of "getting behind", a worker may reach and work ahead on the line. This cheats the recovery time built into the cycle as the product is brought to the worker's position. It also creates an unnecessary reaching and handling time stressor.

Externally paced work often is not adjusted for absences on the line. For example, if four workers share an assembly task, and one worker is absent, then the remaining three workers experience an increase in work load, reduction in cycle time, and a reduction in recovery time if the machine pace is not reduced to compensate for the absence.

Breaks

When tied into incentive pay, employees notoriously work through their breaks, in an attempt to have an additional 30 to 60 min of "productive" time. However, many of the same employees will complain of feeling over-fatigued by midshift. This is related to a lack of movement and appropriate recovery time at the appropriate time. Breaks allow recovery time for the working muscles and offer an opportunity to relax, assume a different posture, move around and restore circulation, eat and provide nutrition to the body.

Breaks are usually factored in by industrial engineers when establishing production standards. Fatigue can occur when the breaks are ignored by workers or unenforced by managers. The absence of an additional break in overtime schedules may also produce excessive fatigue. When production standards are based on an 8-h shift but the production time lasts for 10 h, the absence of at least one more break may result in muscular fatigue and error. Continuous and highly repetitive operations often require a 5 min break for other activities every hour. If this is absent, such as in word processing, overuse and concentrated efforts may result, creating a CTD.

Physical Agents

Vibration

Increased grip forces are needed to hold and steady a vibrating tool or object. Additionally, "vibration white finger" a circulatory disorder in the hand, results from direct exposure to vibration at certain frequencies. The International Standards Organization (ISO) has published guidelines for limiting the amount of continuous hand and arm vibration, setting exposure limits to various frequencies.[23] The effect of vibration on the circulation and nervous system is directly related to the grip force applied to the vibrating object. Higher grip forces impose more vibration onto the hand; lower grip forces create less vibration exposure on the hand.[24] Therefore, vibrating tools, which require higher grip forces, actually create a vicious circle of effects, creating the potential not only for musculotendinous fatigue in the hand, but also vibration white finger which is indirectly related to muscle use.

Vibrating power hand tools are common in today's automated manufacturing plant. The furniture industry is investigating the effects of vibrating sanding equipment on the development of CTDs. A general trend of carpal tunnel syndrome incidence has been noted with the use of hand sanders, which require the hand to rest on a domed control while moving the sander back and forth across the furniture's surface.

Lighting

In general, the lack of task lighting and adequate ambient light will force the head and shoulders into extreme postures as the worker must lean into the worksurface to visually inspect or perform the work.

Temperature

Ambient temperatures less than 50°F result in reduced finger dexterity.[23] Reduced dexterity increases fumbling and inefficient motions. In the food processing industries, temperatures typically must stay below 50°F to ensure shelf life and adequate hygiene. Food products may often be frozen or chilled during manual handling. Touching these products in a chilled ambient atmosphere may doubly affect the fingers. Many workers complain of numb fingers related to the handling of chilled products.

Gloves

Gloves are used frequently in industry to protect the hand for safety or hygiene purposes. However, gloves are a potential source of ergonomic stress. For assembly tasks, both glove size and material have been found to be important in grip forces as well as performance.[25, 26] Investigators feel that glove material is a factor of concern when the task requires maximum grip forces, which is probably associated with the coefficient of friction (cotton vs rubber, for example). Glove size appears to be related more to performance time (improved fit results in maximum performance) rather than grip strength.

The gloved hand, in general, places an additional layer of material between the fingers and the object held. The fingers are supplied with pressure sensitive nerve endings that allow an efficient use of grip so as not to over or under grasp an object; thus, grip force is appropriate to object weight. The additional layer of gloves may mask this ability and result in greater grip forces.

THE PRODUCT ITSELF

Many products inherently create ergonomic stressors due to their size, weight, or special handling needs. We see this in a variety of industries, such as pharmaceuticals, electronics, textiles, apparel manufacturing, red meat, and poultry. When sewing a garment, pinching is required to align the two plies to match patterns or during sewing tasks. The contour of bones and the nonuniformity in animals, and high sanitation standards, create difficulty in developing mechanized means of processing meat and poultry. Some manufacturers' reputations may be built on their high quality standards which require extensive hand work but could result in CTDs. Examples include hand stitching for men's suits, repairing scratches on metal or wood, and inspecting meat products for bone chips. Quality controls that require 100% inspection vastly increase the manual material handling in an operation.

SUMMARY

From this limited discussion of the various potential "root causes" of ergonomic stress, it becomes apparent that no one workplace feature is associated with the development of a CTD, unless its effect is absolutely blatant. Care must be taken to maintain an objective view of the operation, avoid indicting the most obvious stressors and investigate the situation for all potential stressors contributing to the development of CTDs. It may be found that the absolute root cause cannot be addressed without significant modifications to the manufacturing process. However, if other contributing factors are identified and addressed, then the overall level of fatigue and stress will be reduced, even if the absolute root cause cannot be directly eliminated.

SUMMARY OF COMMON "ROOT CAUSES" OF ERGONOMIC STRESS

WORKPLACE LAYOUT AND DESIGN

1. Seating
   - Lack of adjustability
   - Seat pan
   - Back rest
   - Padding
   - Footrests
   - Arm rests
   - Type of seating — traditional chairs vs sit/lean stools
2. Work surface
   - Height
   - Obstacles
   - Orientation
   - Dimensions
3. Workspace
   - Leg room
   - Flooring/footrests
   - Material handling
4. Machine
   - Displays and controls
   - Speed
5. Tooling
   - Effort required to operate
   - Posture required to operate

Handles
Controls

THE JOB DESIGN

1. Methods and training
    Worker technique
    Productivity expectations
2. Training program
    Standard procedures
    Trainers
    Discipline issues
    Inappropriate use of light duty
3. Task and activity mix
4. Overtime
5. Hours per week
6. Standards/incentive pay
7. Machine pacing
8. Breaks

PHYSICAL AGENTS

1. Vibration
2. Lighting
3. Temperature
4. Gloves

REFERENCES

1. Sharkey, B. J., *Physiology of Fitness,* (Champagne, IL: Human Kinetic Publishers, 1979), p. 269.
2. Leveau, B., *Biomechanics of Human Motion,* (Philadelphia: W. B. Saunders, 1977), p. 1.
3. Brunnstrom, S., *Clinical Kinesiology,* (Philadelphia:F. A. Davis, 1972), p. 37.
4. Chaffin, D. B. and Andersson, G., *Occupational Biomechanics,* (New York: John Wiley & Sons, 1984), pp. 37, 39.
5. Rodgers, S. H., "Recovery time needs for repetitive work," *Seminars in Occupational Medicine,* 2(1):19–24 (1987).
6. Laubach, L. L., "Comparative muscular strength of men and women: a review of the literature," *Aviat., Space Environ. Med.* May:534–543 (1976).
7. Mathiowetz, V., Kashman, N., Volland, G., Weber, K., Dowe, M., and Rodgers, S., "Grip and pinch strength: normative data for adults," *Arch. Phys. Med. Rehabil.,* 66:69–72 (1985).
8. Ćatović, A., Kosovel, Z., Ćatović, E., and Muftić, O., "A comparative investigation of the influence of certain arm positions on hand pinch grips in the standing and sitting positions of dentists," *Appl. Ergon.,* June:109–113 (1989).
9. Wiker, S. S., "Effects of Relative Hand Location Upon Movement, Time and Fatigue," Doctoral Dissertation (Ann Arbor: The University of Michigan, 1986).
10. Jonsson, B. and Hagberg, M., "The effect of different working heights on the deltoid muscle: a preliminary methodological study," *Scand. J. Rehabil. Med.,* 3:26–32 (1974).
11. Bjelle, E. A., Hagberg, M., and Michaelson, G., "Occupational and individual factors in acute shoulder — neck disorders among industrial workers," *Br. J. Ind. Med.,* 38:356–363 (1981).
12. De Lacerda, S. G., "Shoulder girdle myofascial syndrome," *Occupational Health and Safety,* (November 1982), pp. 45–46.
13. Hagberg, M., "Workload and fatigue in repetitive arm elevations," *Ergonomics,* 27(7):543–555 (1981a).
14. Hagberg, M., "Electromyographic signs of shoulder muscular fatigue in two elevated arm positions," *Am. J. Phys. Med.,* 60(3):111–121 (1981).
15. Herberts, P., Kadefors, R., and Broman, H., "Arm positioning in manual tasks: an electromyographic study of localized muscle fatigue," *Ergonomics,* 23(7):655–665 (1980).
16. Herberts, P., Kadefors, R., Andersson, G., and Petersèn, I., "Shoulder pain in industry: an epidemiological study on welders," *Acta Orthop. Scand.,* 52:299–306 (1981).
17. Silverstein, B. A., "The Prevalence of Upper Extremity Cumulative Trauma Disorders in Industry," Doctoral Dissertation, (Ann Arbor: The University of Michigan, 1985).
18. Rodgers, S. H., "Job evaluation in worker sickness determination," *Occupational Medicine:State of the Art Reviews,* 3(2):219–239 (1988).

19. Kroemer, K. H. E., "Cumulative trauma disorders:their recognition and ergonomics measures to avoid them," *Appl. Ergon.*, 20(4):274–280 (1989).
20. Chaffin, D. B. and Greenberg, L., *Workers and Their Tools: A Guide to the Ergonomic Design of Hand Tool and Small Presses*, (Midland, Michigan: Pendell Publishing Co., 1977).
21. Fraser, T. M., *Ergonomic Principles in the Design of Hand Tools*, (Geneva, Switzerland: ILO, 1980).
22. Eastman Kodak Company, *Ergonomic Design of People at Work, Vol. 2*, (New York: Van Nostrand Reinhold Company, 1986), Chapter 15.
23. ACGIH, *Threshold Limit Values for Chemical Substances and Physical Agents*, (American Conference of Governmental Industrial Hygienists, 1990) p. 74.
24. Wilson, J. R. and Corlett, E. N., *Evaluation of Human Work*, (Philadelphia: Taylor and Frances, 1990) p. 446.
25. Chen, Y., Cochran, D. J., Vishu, R. R., and Riley, M. W., "Glove size and material effects on task performance," in *Proceeding of the Human Factors Society 33rd Annual Meeting*, (Santa Monica, CA: The Human Factors Society, 1989), pp. 708–712.
26. Cochran, D. J., Howe, E. R., and Riley, M. W., "An evaluation of commercially available plastic handled knife handles," in *Proceeding of the Human Factors Society 29th Annual Meeting*, (Santa Monica, CA: The Human Factors Society, 1985), pp. 802–806.

ADDITIONAL READING

"Diverse workforce posing new challenges in designing ergonomic workplace," *Occupational Safety and Health Reporter,* The Bureau of National Affairs, Inc., 0095-3237/90 p. 995.

Drillis, R. and Contini, R., *Body Segment Parameters,* B. P. 174–942, technical report #1166.03, School of Engineering and Science, New York University, New York, 1966.

May, V. R., "Analyses of Work Functioning Data From a Work Capacity Evaluation Program," Paper presented at the Virginia Orthopaedic Association Annual Spring Conference (Williamsburg, VA: May 1, 1987).

"Repetitive Motion Injuries Cause Gain in Job Illnesses," in *Greensboro News & Record*, November 15, 1990.

U.S. Bureau of Labor Statistics, *Employment and Earnings,* 1987.

Chapter 3
The Disorders and Their Etiology

Overuse syndrome was probably first described by Hippocrates around the year 400 B.C. when he warned athletes about the risk of sustained maximum performance.[1] About 200 years ago, Ramazzini described the perils of violent and irregular motions.[2] In the 1800s, a number of articles appeared in the medical literature pertaining to overuse syndromes in musicians and others. These conditions were variously described as musician's cramp,[3] pianist's cramp,[4] writer's cramp,[5] and treadler's cramp.[6, 7] More recently, Fry described overuse syndrome in musicians.[8] The most recent terminology for soft tissue overuse, advocated by the National Institute of Occupational Safety and Health, is CTD.[9] Other terminology includes repetitive strain injury, repetitive motion syndrome, and overuse syndrome. Whatever the name, they are all quite similar and are nothing new.

These conditions have become exceedingly common in modern industry, and they are not limited to the U.S. Australia[10] had a devastating epidemic of these problems in the 1980s. Reasons for the rapid increase are probably numerous. These include changes in production technology, the advent of more fine hand work and less heavy muscular work, and increased utilization of finished products, such as chicken and meat fillets. Awareness has also increased on the part of government, doctors, and employees. The current legal climate and workers' compensation system also contribute to the increase in the incidence of CTDs. Many of these conditions were probably treated as nonoccupational illnesses in times past.

In the U.S., a popular belief is that the majority of CTDs are tendon-related disorders resulting from overuse,[9, 11] manifested as an acute or chronic tendonitis, median nerve compression, or inflammation at the wrist resulting in carpal tunnel syndrome. Another concept has not received adequate attention in the U.S. This concept states that these conditions are primarily muscular, resulting from overuse in the relatively new or unconditioned employee; and, for reasons not fully understood, CTDs may occur in employees who have been doing the same job without difficulty for longer periods.

In the latter case, the overuse syndromes which develop many years after rehabilitation from poliomyelitis could perhaps provide some insight.[12] When chronic, these conditions behave like fibromyalgia or fibrositis except that they are localized to the area that has been overstressed. They may spread to other areas in the upper extremity; once chronic, they become very difficult to treat. Usually one can observe tender points (trigger points of Smythe[13]) in the superior border of the trapezius, in the scapular region, supraclavicular region, and quite often in the flexor or extensor regions of the forearm. As with fibromyalgia, these are often accompanied by numbness or paresthesia in the hand, forearm, or upper arm.[14] The condition was well described by Miller and Topliss[10] in a series of 229 cases of overuse syndrome in Australia. Muscle biopsies[15] have revealed changes in the affected muscle; however, these changes can also be seen in asymptomatic muscle.

Very likely, both concepts may explain some of the CTD cases experienced today. Hadler[16] believes that CTDs are an iatrogenic concept. Many believe that CTD severity is strongly influenced by the concern that these conditions cause among workers.

In the Australian series cited above, only 1 out of 229 overuse syndrome cases was diagnosed as carpal tunnel syndrome. However, in the U.S., CTDs seem to have become almost synonymous with carpal tunnel syndrome in the minds of many physicians and the lay public.

CTDs encompass a variety of conditions. The type of structure affected and its location creates the basis for naming the disorders. They may be acute, subacute, or chronic. They may be of unknown cause, secondary to some disease process (such as hypothyroidism), or caused by repetitive work. Work may not be the primary cause of these conditions, but may aggravate them.

One common mistake is to think of the conditions independently. Actually, many people suffering from CTDs have relatively large areas of the body affected. Therefore, we should think of these as upper quarter disorders which may simultaneously affect the neck, shoulder, elbow, forearm, wrist, and hand.

MYALGIA (MUSCLE PAIN) AND MYOFASCIAL SYNDROMES

Myalgia simply translates as "pain in the muscle". The conditioning process and the pathogenesis of muscle pain have been previously discussed. Muscle pain is exceedingly common. In a survey of a large textile company,[17] 47% of employees indicated that they currently had some type of problem with their upper extremity. These problems were not limited to jobs requiring repetitive or forceful motions. Those nonrepetitive jobs also had problems but to a lesser extent.

Muscle Pain During Break-in on a New Job

Muscle pain is often experienced by almost every new worker who has to do any type of unaccustomed strenuous or repetitive work. It is sometimes referred to as "break-in soreness" or "green hands". It is often accompanied by some degree of spasm and difficulty in performing agile body movements. Palpation of the affected area will often reveal tenderness and spasm. This type of muscle soreness is usually self-limiting. If the work is a one-time task, (e.g., a weekend endeavor or working in a warehouse for a day) cessation of the work will usually result in relief within several days. Even though this type of soreness is self-limiting and benign, it probably is accompanied by some degree of muscle damage which soon heals. Experience indicates that a gradual break-in can result in conditioning with a minimum of soreness.

Usually the soreness clears due to the conditioning process even though the same work is continued. The muscle gradually heals over a period of several days and becomes accustomed to the activity. However, once conditioned, the muscle will rapidly become deconditioned after the activity has stopped for as little as 1 week. One month of inactivity can result in significant deconditioning and can require another conditioning period after return to work. Employers and employees should be aware of this since long vacations, and absences after illnesses or accidents, will result in the need to have the employee reconditioned. Personal leaves and pregnancy leaves are other examples where deconditioning can occur.

Subacute and Chronic Myalgia

One of the signs of benign myalgia during break-in is that it is usually relatively diffuse over the affected body region. It improves rapidly over a period of 1 to 2 weeks even though the activity is continued. Alternately, subacute chronic myalgia does not spontaneously improve. It frequently localizes in certain groups of muscles. At this point it is often called "myofascial pain syndrome". Careful palpation of the area of complaint will usually reveal tenderness and considerable firmness as compared with the same muscle on the opposite unaffected side, or adjacent muscles. In CTDs there may be pain and spasm along the superior border of the trapezius, the muscles between the scapulae, the brachioradialis muscle region, and the flexor or the extensor muscles of the forearm. Pain is usually elicited by stretching of the muscle; for example, extension at the wrist in the case of forearm muscle spasm. The condition usually resolves with appropriate therapy, such as heat or cold, stretching exercises, and perhaps some reduction of activity.

A common mistake, however, is to believe that it is only a painful spastic muscle and that it will spontaneously resolve. The pain tends to result in the spasm, and the spasm results in more pain causing a "pain-spasm" cycle. This cycle will usually not be relieved without appropriate therapy.

Chronic Myofascial Pain Syndromes

A perplexing enigma in CTDs exists when a seemingly simple myalgia may progress to a chronic myofascial pain syndrome with muscle spasm, dysfunction, and eventually, temporary or permanent disability. In the chronic stages, affected individuals have many of the characteristics of fibromyalgia or fibrositis syndrome.

Fibromyalgia or fibrositis is characterized by chronically painful spastic muscles. It may be aggravated by rest and be worse upon first arising in the morning. It may also be aggravated by repetitive motion. Fibromyalgia is frequently accompanied by paresthesia and may also be accompanied by other bodily symptoms such as lack of sleep, nervousness, etc. Fibromyalgia may be classified as "primary" or idiopathic, e.g., not having any known cause; or "secondary", due to other disease processes. A characteristic of fibromyalgia syndrome is the presence of trigger points or tender points. These are small areas of tenderness, usually about the size of the end of the examining finger, surrounded by nontender muscle. Frequently the patient is unaware of these trigger points. Palpation of them will often result in pain which shoots down or up the extremity, often duplicating the symptoms of which the patient complains. Trigger points have been described in detail by Smythe.[13] Travell and Simons[18] have written an entire book about their occurrence and treatment.

Fibromyalgia can be treated by injections into the trigger points, and cooling with fluorocarbon sprays, followed by stretching. This will often provide relief. Injecting them with lidocaine or procaine may also provide relief. However, the overall condition is quite resistant to treatment.

Localized pain in a muscle may persist for weeks. It is usually readily diagnosable and treatable. On the other hand, the fibromyalgia syndrome is much more chronic, resistant to treatment, tends to involve multiple areas, and is accompanied by other symptoms. A secondary fibromyalgia syndrome from overuse may be the difference between an employee with a benign, painful spastic muscle and one who is virtually disabled and complains of continuous pain long after the offending cause has been modified. In the Australian series of 229 patients, the author found several tender points in the arms of most of the patients. In biopsy studies, Frye has found characteristic muscle lesions, which have also been described in fibromyalgia syndrome. On the other hand, the fibromyalgia syndrome itself is controversial and some believe is an iatrogenic concept. They believe the condition is fueled by the physician's desire to name and treat something.

Certainly the overuse syndrome, described in musicians and others in earlier and recent literature, appears to be similar to the fibromyalgia syndromes described above. What is it that causes a person to progress from having a painful sore muscle to a fibromyalgia type of illness? The physical reasons are poorly understood.

Other complicating factors are

1. Symptom exaggeration in hypersensitive, nervous, and anxious patients
2. An angry patient who has been caught in a conflict between management and the legal system
3. Secondary gain from workers' compensation
4. The reinforcement of symptoms by physicians that label and treat a disorder
5. A legal climate which encourages litigation
6. Regulatory and media attention to the problem
7. The minimally studied involvement of the body's connective tissues that surround the affected muscles

Looking for any or all of these to explain the broad spectrum of chronic myofascial pain syndromes would be an oversimplification. Many musicians are highly motivated and accomplished; yet they appear to be quite prone to these conditions. These conditions have occurred for many years, long before worker's compensation and notions of legal liability were in vogue. These conditions also often occur in well-motivated employees who have worked satisfactorily on the same job for years without difficulty.

One recent case of particular interest to this author, involved a middle-aged male who had done a job called "three-pointing" in which the thighs of a turkey are lifted upward and placed into a shackle. This involves repetitive grasping and exertions with the hands and forearms. After 14 years without symptoms, the employee developed exquisitely painful spastic forearm muscles. He also experienced paresthesia in the hand. He was removed from the job and placed on modified work, but without improvement. He became hostile and aggravated and pleaded that he just wanted to go back to the same job which he said he dearly loved. Surgeons wished to operate on him for carpal tunnel syndrome, but clinical symptomatology was not characteristic. His nerve conduction studies were normal. He was finally referred to a physiatrist and made gradual improvement over a period of time, but was not able to return to that job. Eventually he was operated upon, developed complications of infection, and did not improve.

Though there is considerable controversy over the true nature of work-related myofascial pain syndromes, few argue that these people have discomfort and pain. Therefore, early care aimed at alleviating symptoms, and correcting predisposing activities is appropriate.

The Importance of Myalgia

It has been obvious for many years that myalgia and myofascial pain syndromes can cause pain. Clinical experience reveals that when there is progression to a more serious state, early warning signs are often not heeded. As the condition progresses, providers of treatment may not recognize the true nature of the problem. Hostility and anger develop, symptoms become exaggerated, and then far more serious problems arise. At this point surgery may be performed, but frequently provides little relief and may even aggravate the condition.

Muscle conditions can predispose the development of tendon problems. Histories of many people that have developed tendonitis reveal that frequently pain and muscle spasm preceded the tendonitis.

The development of carpal tunnel syndrome may be related to the stress and compression upon the tendons as they pass through the carpal tunnel. Spastic muscles exert more force and friction upon the tendons than well functioning relaxed muscles. Therefore, early recognition, treatment, and care of persistent myalgia and myofascial problems is of utmost importance in preventing carpal tunnel syndrome.

TENDONITIS AND TENOSYNOVITIS

A tendon may be bare or pass through a sheath. The sheath protects, guides, and lubricates the tendon in places where there is a great deal of movement or potential friction. Through much of their course, tendons at the wrist, shoulder, hand, foot, and ankle are surrounded by tendon sheaths. Tendons at the elbow, knee, and forearm are without sheaths. When a tendon that is enclosed in a sheath becomes inflamed, the condition is called "tenosynovitis". When a tendon without a sheath becomes inflamed, it is called "tendonitis". Another medical term for the latter is enthesopathy, meaning pathology, at the insertion of a tendon. Where tendons insert into the bone there is a transition between tendon and bony tissues. Enthesopathy refers to a disorder at these insertions. Enthesopathy is at the lateral elbow which is referred to as "tennis elbow" or "epicondylitis".

In terms of symptoms and early treatment, these conditions are quite similar. However, in more chronic stages, when there is a sheath surrounding the tendon, inflammation is more likely to result in "triggering". When this happens the tendon becomes trapped or temporarily hung up in the sheath, resulting in difficulty in movement of the digit of that tendon.

Causes of Tendonitis

Tendons have great tensile strength and can tolerate strong pulling without harm. However, when the pull on the tendon exceeds the strength of it or its bony or muscular attachment, injury occurs as a complete or partial rupture. Tendons are commonly pulled, stretched, and squeezed repeatedly during work. In addition, bending and kinking of the tendons create mechanical friction stress. This may result in fraying of the tendon fibers. Inflammation occurs in the tendon, its surrounding sheath, or in the surrounding tissues when there is no sheath. Edema may be present within the tendon sheath creating additional pressure and resistance.

Acute tendonitis is very painful; this pain is intensified by passive or active movement of it. Palpation of the tendon also elicits tenderness and there may be localized swelling over it. Another relative common finding is crepitation. With crepitation there is the sensation of rubbing and crackling when the tendon is moved.

Other Causes of Tendonitis

Tendonitis can also occur as a result of nonwork-related problems, such as in rheumatoid arthritis and certain types of osteoarthritis, especially where the tendon courses over the roughened bone or joint. Gout may also cause tendonitis. Another form, calcific tendonitis, may be due to a metabolic disorder or may follow injuries to the tendon after which calcium is deposited.

Prognosis and Treatment

Acute tendonitis of any significant severity is usually a very painful condition and the tendon must be immobilized. This can be accomplished by splinting. In addition, local ice massage, or injection with local anesthetics and corticosteroids are often helpful. Aspirin, acetaminophen, or nonsteroidal anti-inflammatory agents help to reduce the pain and inflammation. Acute tendonitis, if properly treated, usually clears rather promptly. However, care must be taken not to overuse the tendon again. Even more importantly, activities should be gradually resumed so as not to cause a flare-up of the condition.

If the tendonitis does not soon resolve completely it may become subacute or chronic. This becomes a much more difficult condition to treat. There may be repeated flare-ups with overactivity. These flare-ups may be controlled with ice massage, gentle stretching of the muscles attached to the tendon, and deep friction massage. The latter is accomplished by stroking the tendon at a right angle with the index finger. This tends to rearrange the disorganized fibrils and assists in healing. When a sheath becomes narrowed resulting in triggering, surgical release may be necessary. In all cases of tendonitis, ergonomic intervention to evaluate biomechanic and design problems is indicated.

Specific Types of Tendonitis
Shoulder

Tendonitis in the shoulder region is a common occurrence in people who must lift continuously or work overhead. It is also common in athletes who must forcibly throw an object such as a baseball or football. Tendons commonly involved at the shoulder attach to the biceps, in which case it is called bicipital tendonitis. This involves the long head of the biceps tendon as it courses over the head of the humerus through the bicipital groove. In bicipital tendonitis, pain results from stretching of the tendon by straightening the elbow, hyperabducting the elbow or arm, or by lifting weights with the biceps. The tendon is also tender to palpation. Rotator cuff tendonitis results when the tendons of the muscles which make up the rotator cuff become inflamed. Rotator cuff tendonitis will cause pain when the shoulder is outwardly rotated, or with throwing motions.

Epicondylitis

The most common form is lateral epicondylitis, also known as tennis elbow. The muscles which extend the wrist and supinate the hand attach in a short tendon to the lateral epicondyle of the humerus. The region of this attachment is a common site of inflammation in those who do repetitive, forceful twisting motions; for example, frequent rotary motions of the wrist, with a screwdriver, or frequent lifting with the palm down. Force upon this tendon is also exerted in a tennis player using his back hand, thus the term "tennis elbow".

When the tendon at the medial epicondyle is inflamed it is called medical epicondylitis, sometimes known as "golfers elbow". This is caused by forceful pronations of the forearm or forceful repeated flexion at the wrist. Both medial and lateral epicondylitis can be caused by a bump up on the elbow which more commonly occurs at the lateral epicondyle.

The principles of treatment of tendonitis also are applicable to epicondylitis. Included is ice massage in the acute phases. The employee is cautioned about the motions which aggravate the condition, twisting of the hand and picking things up with the palm down. A small tennis elbow band may be used just beneath the elbow to provide a certain amount of support for the involved muscles. Injection of local anaesthetic with corticosteroids may also be helpful, but generally should not be repeated more than once or twice. Epicondylitis is often resistant to treatment, and tends to recur frequently. However, in most cases, it can be controlled so that recurrences are infrequent and the chronic condition is mild. The patient is best cautioned about this so that he/she does not expect to be fully relieved of symptoms within a short period of time. Rarely should surgery be necessary for these conditions.

de Quervain's Tendonitis

At the radial side of the wrist, the tendons of the abductor pollicis longus and the extensor pollicis brevis converge and pass through a common sheath, inserting into the metacarpal and first phalange of the thumb respectively. Since the thumb is involved in so many repetitive motion jobs, this tendon at the wrist is a common site of tenosynovitis. Stretching or pulling on the tendon by passive or active flexion or abduction of the thumb reproduces the pain as does palpation directly over the site. There may be swelling and crepitation over the area. General principles of treatment of tendonitis apply. In acute phases, the thumb should be splinted. A common mistake in treating this tendonitis is that of splinting only the wrist and leaving the thumb free. The thumb must be tethered in a thumb spica splint in order to be immobilized. If the tendonitis is recurrent and results in stenosis of the sheath, surgical treatment may be necessary, and is usually effective.

Tendonitis about the Wrist and Hands

Any of the flexor or extensor tendons in this region may develop tendonitis. Extensor tendons are commonly involved where there is frequent stretching, flexion, and grasping. It is not uncommon to see swelling along the sheaths of the involved tendons. Flexor tendonitis, if it occurs where the flexor tendons pass through the carpal tunnel, may be accompanied by swelling of the tendon sheath and compression of the median nerve resulting in carpal tunnel syndrome which will be described below.

Trigger Finger

We have already mentioned "triggering" which is the partial or complete entrapment of the tendon within the sheath. This is often seen in those who must squeeze a tool causing pressure on the palm of the hand while there is stretching and pulling upon the tendon. The tendon and its sheath become inflamed. With prolonged inflammation stenosis of the sheath occurs. A nodule may also form on the tendon, frequently at the proximal end of the tendon sheath. Then flexion of the digit causes the nodule to pass through and outside the sheath when the finger is subsequently extended. The nodule is caught at the narrowed end of the sheath resulting in movement difficulty and a cracking, popping sensation when re-extending the finger. This commonly occurs at the index finger, thus the term "trigger finger". Trigger finger is similar to other types of tendonitis. This type of tendonitis must be treated early if surgery is to be avoided. Typically, in the early stages the worker will complain of these symptoms at the end of the day or after using a particular tool. This is an early warning sign. Treatment, including local massage and stretching exercises of the flexor muscle and tendons, will often be effective at this stage. Local injection of the sheath or nodule may also be helpful. With early treatment, ergonomic evaluation, and stretching exercises to open the palm and stretch the fingers, surgery can often be avoided. Surgical treatment involves opening the sheath and/or removal of the nodule that is causing the difficulty.

ENTRAPMENT SYNDROMES

Most entrapment syndromes involve nerves; several may involve blood vessels as well. Entrapment syndromes may be due to diseases such as arthritis, hyperthyroidism, pregnancy, or a wide variety of collagen vascular disorders or edema of any cause. Quite often, however, they are caused by temporary or readily remediable conditions such as muscle spasm, muscle hypertrophy, or muscle relaxation due to poor physical condition or aging. Entrapment may occur between two muscles, between two bones, between bone and ligament, between ligamentous structures, or at the intervertebral foramina. Entrapment syndromes usually cause symptoms distal to the lesion, but they also may produce retrograde symptoms proximal to the lesion making it difficult to localize.

There are features of CTDs which add to the complexity of dealing with entrapment syndromes. The disorders often involve multiple sites in the upper extremity. Dealing with only a single lesion may be futile. If the entrapment occurs in a proximal region; for example, neck or shoulder, then the distal part of the nerve may be more susceptible to injury and a subsequent entrapment syndrome. For example, impingement of the nerve roots as they exit the intervertebral spaces of the neck is a common complicating factor in carpal tunnel syndrome. This is called the "double crush syndrome" which means that the nerve is impinged at more than one spot.[19] For this reason, it is very important for all concerned — ergonomists, plant engineers, supervisors, nurses, and physicians — to consider the whole upper extremity, and in fact, the upper back when considering CTDs.

Cervical Radiculopathy

The discussion of anatomy indicated how the nerve roots exit the spinal canal at the intervertebral foramina in the neck. The cervical vertebra are some of the most frequently moved joints in the body and accordingly, are subjected to wear which may result in degenerative disk disease, or arthritis. Arthritis can cause impingement upon the foramina. This then can cause numbness, parethesias, or weakness in the upper extremity depending upon the site of the impingement.

Although diagnosis of cervical arthritis and disc degeneration is not necessarily difficult with modern imaging techniques, it is often difficult to determine if these are the cause of the problem. Just because one has cervical arthritis demonstrated by X-rays does not necessarily mean that this is the cause of the discomfort. A trial of therapy, including stretching exercises and nonsteroidal, anti-inflammatory drugs may be helpful. Cervical radiculopathy has been cited[20] as the most common cause of failure of surgery for carpal tunnel syndrome. Therefore, it is important to address this situation, if it exists, prior to consideration of surgery.

Thoracic Outlet Syndrome

Thoracic outlet syndrome may be caused by anatomical anomaly, but this is present in only about 20% of the more serious cases requiring surgery.[21] Most of the cases are due to poor muscle condition,[22, 23] or spastic muscles in and about the region of the shoulder area. Sagging muscles, heavy breasts, and tight fitting brassieres, can also contribute to this syndrome. Observing the work habits of the individual provides a tip-off that thoracic outlet syndrome may be occurring. If the job involves considerable overhead work, then the muscles of the shoulder and neck region tend to become tight and may cause the compression. Thoracic outlet syndrome is relatively common in musicians.[24] The neurovascular bundle, as it traverses the cervico-axillary canal, may be entrapped at one of three areas: (1) between the scalenus muscles in the neck, (2) between the clavicle and the first rib, and (3) between the chest wall and the pectoralis minor muscle. Also, a cervical rib may be present which contributes to the entrapment.

Symptoms include a sense of discomfort and fullness in the arm and numbness and parethesias in the arm and the hand. Usually parethesias will involve the ulnar region, but also may involve other parts of the hand and forearm. Often the patient complains of numbness and tingling in the whole hand.

Symptoms of thoracic outlet syndrome can be demonstrated in a normal individual by hyperabducting and internally rotating the shoulder (placing the arm in back) which will reproduce and accentuate the symptoms.

There are a number of diagnostic maneuvers which physicians can use including the Adson's[25] and Roos[21] sign, and others. Unfortunately, these are often positive in normally asymptomatic people. Listening for a bruit over the thoracic outlet may also be helpful, but is not an uncommon finding in normal people. Doppler tests, arteriograms, and nerve conduction studies may be helpful, but are not diagnostic. Perhaps the best approach is a very careful analysis of the person's symptoms, careful physical exam looking for tender points in the regions that might bring about muscle spasm, and looking for spasm, tightness, or drooping. If these indicate the probability, corrective measures will often bring about quick relief. These include exercises designed to improve the strength of the muscles in the region and stretching to relieve spasm. We have seen many cases scheduled for carpal tunnel surgery who have had their symptoms relieved by treatment for thoracic outlet syndrome.

Cubital Tunnel Syndrome

The ulnar nerve traverses through the cubital tunnel at the medial side of the elbow. Frequent pressure upon the ulnar nerve, or twisting motions at the elbow may result in sliding of the nerve through the tunnel, inflammation, and swelling. Symptoms include pain and soreness at the site of the canal, parethesias and shooting pains into the forearm, and parethesias of the ulnar aspect of the hand which includes the ulnar half of the ring finger and the little finger. Injection with procaine and corticosteroid at the elbow may be helpful. Conservative treatment also includes ice and anti-inflammatories. Surgical treatment is aimed at opening the canal. Sometimes transplant of the ulnar nerve is necessary.

Radial Canal Syndrome

The radial nerve passes just below the lateral epicondyle and into the forearm near the head of the radius. Compression of the radial nerve at this area due to contracted muscle or bones will result in decreased sensation in the distribution of the radial nerve and pain in the region of the lateral epicondyle. The pain is reproduced by passive stretching or resisted extension of the middle finger.

Anterior Interosseus Syndrome

The anterior interosseus branch of the median nerve innervates some of the deep muscles of the front of the forearm and the flexor pollicis longus. It can be compressed by the deep anterior forearm muscles. Causes include acute trauma, and in some cases, overuse. Since the nerve is purely motor, symptoms include pain in front of the forearm near the elbows, which is made worse by repetitive high demand activities. Classically, there is weakness of the thumb/index finger pinch, with a tear drop, rather than an "O" pattern with the pinch.

Posterior Interosseus Syndrome

A deep branch of the radial nerve (interosseous) can be compressed within the supinator muscles in the forearm. Symptoms may be similar to lateral epicondylitis. Weakness of the extensor muscles occur, especially the little finger. The condition may result in decreased sensation over the sensory distribution of the radial nerve.

Forearm Entrapment Syndrome — Pronator Teres Syndrome

The median nerve, as it enters into the forearm, passes beneath the two heads of the pronator teres muscle. When this muscle becomes enlarged or inflamed due to constant pronation, the nerve can become entrapped between these two heads.

Symptoms include tenderness and evidence of spasm and tightness over the pronator muscle. Tapping on this area may produce a Tinel's type of sign, which is a sharp electrical shock-like sensation radiating into the forearm and hand. The pain is reproduced resistant to pronation and flexion of the wrist. Due to the fact that the entrapment is of the median nerve, symptoms will be similar to that of carpal tunnel syndrome and are often mistaken for it. Therefore, it is very important to evaluate the type of motions and also to carefully examine the forearm region. In pronator teres syndrome, the median nerve is entrapped above the wrist. The palmar cutaneous branch of the median nerve branches from it just about the transverse carpal ligament and traverses superficial to the transverse carpal ligament. It provides sensation to the palm of the hand near the wrist, and will be affected in pronator teres syndrome, but not in carpal tunnel syndrome.

Carpal Tunnel Syndrome

The most common nerve entrapment of the upper extremity is the carpal tunnel syndrome. The carpal tunnel is made up of two rows of carpal bones, eight in all, which bound the tunnel posteriorly, medially, and laterally. The transverse carpal ligament bounds the tunnel on the palmar side of the wrist. The median nerve and nine flexor tendons pass through the carpal tunnel. The contents of the carpal tunnel fit snugly — there is little room for expansion of the contents, or narrowing of the tunnel since either will cause compression. Therefore, carpal tunnel syndrome can result from anything which reduces the cross sectional area of the tunnel or anything which causes the contents of the tunnel to swell.

Many disease processes can cause carpal tunnel syndrome. Indeed it is often forgotten that in a significant percentage no cause is known. A list of diseases which may be associated with carpal tunnel syndrome is shown on Table 1. Diabetes, rheumatoid arthritis, and osteoarthritis of the cervical spine and carpal bones are commonly associated. Carpal tunnel syndrome is not an infrequent complication of the third trimester of pregnancy when edema may result in swelling of the contents within the canal. Carpal tunnel syndrome is more common in females than males, with a ratio of about 2 to 1. It often occurs in middle aged females and may have no relationship to activity.

Work Related Activities and Carpal Tunnel Syndrome

Dr. George Phalen,[26] one of the early pioneers, in describing carpal tunnel syndrome stated that it was only rarely an occupational disease. However, he felt it could be aggravated by occupational causes. Others[27] support this viewpoint. However, Armstrong[28] has developed a biomechanical model for carpal tunnel syndrome. Frequent flexion, extension, or ulnar deviation of the wrist, combined with repetitive and/or forceful movements of the digits results in increased friction of

TABLE 1. Nonoccupational Conditions Which Cause or Are Associated with Carpal Tunnel Syndrome

Diabetes
Rheumatoid Arthritis
Osteoarthritis of Carpal Bones
Pregnancy
Trauma
de Quervain's Tendonitis
Ganglia
Previous Fractures
Myxedema
Hyperthyroidism
Acromegaly
Amyloid Disease
Scleroderma
Systemic Lupus Erythrodermatitis
Dermatomyositis
Hunter's Syndrome
Scheie's Syndrome
Cervical Arthritis (Spondylosis)
Raynaud's Disease
Gout
Familia Occurrence

and pressure upon the flexor tendons as they course through the carpal tunnel. This may result in swelling of the sheaths of the tendons. Due to the friction, the median nerve may also begin to swell and flatten. At first these changes may be primarily those of edema or inflammation and may subside rapidly with rest, splinting, injection and anti-inflammatory agents. In time, however, more permanent changes, including scarring and adhesions around the nerve, and thickening of the tendon sheath and of the transverse carpal ligament, may occur. At this stage, the process may become permanent and no longer respond to conservative therapy, and surgery may be necessary. Silverstein et al.,[29] and Armstrong et al.[30] in plant studies, found an association of carpal tunnel syndrome with forceful repetitive jobs. In the past decade, numerous studies have been published associating carpal tunnel syndrome and overuse.[31-38] In spite of different viewpoints, the above hypotheses are predominant amongst clinicians.

Symptoms of carpal tunnel syndrome include pain in the wrist, hand, and forearm. Pain may also shoot up into the arm and shoulder. Numbness and parethesias in the median nerve distribution of the hand are also salient features. The median nerve provides sensory fibers to the first three fingers and one half of the fourth finger. The person affected with carpal tunnel syndrome frequently awakes during the night with numbness, tingling, and pain in this distribution. This is probably due to sleeping with the wrists in a deviated condition and/ or a certain amount of congestion created by lack of activity. Shaking of the hands relieves the symptoms. In time the symptoms progress, are more frequent, and occur during the day.

In time, motor functions may become affected. The muscles of the thenar eminence, which include the opponens and pollicis abductor pollicis brevis, become affected, resulting in weakness of tip pinch grip and abduction of the thumb. The individual may not be able to hold a piece of paper placed between the tip of the thumb and the index finger. Weakness and clumsiness may result and the individual begins to notice dropping of tools and other objects. As muscle atrophy progresses, the thenar eminence may appear flattened. Actually, this latter finding seems to be quite rare now that carpal tunnel syndrome is usually recognized so early. The lumbrical muscles which flex the metacarpal phalangeal joints and extend the two distal phalanges may also become weakened.

Examination of the hand reveals decreased sensation in the areas described. The ability to discriminate two points at the fingertips may be reduced so that the individual can only distinguish them from a single point when they are at least 5 mm apart. Carpal tunnel syndrome has been staged into early, intermediate, and advanced stages.[39]

Nerve conduction studies are considered the gold standard by some, but of not great value by others. In these studies, the median nerve above the wrist receives an electrical shock. A transducer placed upon the thumb records the time that it takes for the shock to travel from the site above the wrist and result in muscle contraction at the thumb. Therefore, this interval consists of both the nerve conduction time and the time for the muscle to respond, and is called the distal motor latency. Motor latency in normal people is normally at or below 4.6 msec.[40]

Another type of latency is sensory which is determined by placing an electrical shock on the skin of one of the digits, typically the middle finger and recording the amount of time that it takes for the impulse to reach the median nerve above the wrist. Normally, this latency is 3.5 msec or less.[40] Usually, the sensory latencies are the first that are delayed in carpal tunnel syndrome. In approximately 10% of the cases of carpal tunnel syndrome, both motor and sensory latencies are normal. In addition, these studies may be falsely positive. Temperature changes and failure to warm the limb to at least room temperature can cause false positives. A prolonged latency can result from a proximal lesion of the nerve, and therefore, does not localize the lesion at the carpal tunnel. In order to do this, nerve conduction velocities should be performed, wherein the nerve conduction velocity in the forearm and also across the wrist is measured. This technique has been described by Kimura.[41] In a series of 639 extremities with presumptive cases of carpal tunnel syndrome, they found that there was a decrease of 25% or more of median sensory nerve conduction velocity across the wrist as compared with the forearm. Various types of neuropathies can also result in prolonged latencies.

One would think that with a lesion so sharply localized as the median nerve lesion in carpal tunnel syndrome, that it would be relatively easy to diagnose and treat. In cases of carpal tunnel syndrome of unknown cause or those associated with pregnancy or disease state, the diagnosis is often clear-cut and the symptoms are quite typical. Even in the absence of positive nerve conduction studies, the surgeon may not hesitate to operate because of the classical symptoms. Unfortunately, carpal tunnel syndrome associated with overuse is quite a different situation. As we have already mentioned, there is frequently associated muscle inflammation or other entrapments in the upper extremity. Frequently the carpal tunnel syndrome is mild and symptoms and findings are equivocal while symptoms of other conditions predominate. In such cases, it is very important to provide conservative treatment and also to evaluate and treat vigorously other associated problems such as forearm muscle spasm or thoracic outlet syndrome. Otherwise surgical treatment will fail, as is often the case.

The treatment for carpal tunnel syndrome may be conservative or surgical, depending upon its stage. Conservative treatment consists of anti-inflammatory drugs and ergonomic evaluation of the job. The latter will include biomechanic analysis of the various wrist motions. Splinting of the wrist may be helpful, however, one must be cautious with this. It may result in raising the elbows during work, which may result in painful elbow or shoulder problems such as the thoracic outlet syndrome. Injection of the wrist may provide temporary relief in many cases and permanent relief in over half according to some series.[42] Pyridoxine (vitamin B_6) has been used. Results are controversial — some observers reporting success and others no help.[43–49]

Entrapment of the Ulnar Nerve at Guyon's Canal

The ulnar nerve passes into the hand at Guyon's canal on the medial side of the wrist. It is outside of the carpal tunnel. This canal is not so subject to entrapment as the carpal tunnel due to the fact that there are no tendons traversing the canal. Entrapment of the ulnar nerve at Guyon's canal results in parethesias, numbness, and tingling of the little finger and part of the ring finger.

VIBRATION WHITE FINGER

Frequent vibration can cause injury to the small blood vessels in the hand. This type of injury has been seen relatively frequently in the past, before modern dampening techniques in such workers as chainsaw and jack hammer operators. The injury to the blood vessels causes a lack of circulation resulting in blanching or whiteness of the fingers. Feeling the fingers reveals them to be cold. Elevation of the hand results in blanching of the fingers and failure of color to return quickly after the hand is lowered. Accompanying the blood vessel changes may be injuries to the small nerves, also resulting in feelings of numbness and coldness similar to that which may occur with the nerve entrapments. Modern diagnostic techniques, doppler and arteriography, are helpful in evaluating these syndromes.

Vibration white finger may also be confused with Raynaud's disease with which it is similar. Actually, the phenomenon caused by vibration white finger is essentially the same pathology as Raynaud's disease. The presence of risk factors, such as vibration, helps to distinguish one from the other.

Another microvascular disease, Berger's disease, is also associated with blood vessel abnormalities and may result in blanching. However, Berger's disease typically involves the larger blood vessels. It is strongly associated with smoking.

ARTHRITIS

Arthritis of the hand and wrist is common. It also is a disease frequently seen in middle aged and older females. There are two major types of arthritis: (1) inflammatory arthritis, which is a generalized process associated with diseases such as gout or rheumatoid arthritis, and (2) osteoarthritis which is a degenerative joint disease. On the continuum, there are variants such as inflammatory osteoarthritis of the hand which have features of both diseases.

Degenerative joint disease or osteoarthritis can be associated with the "wear and tear" of continuous use, but recent findings also indicate that there is a genetic predisposition to it.[50] The question of whether significant disabling arthritis is associated with repetitive work is controversial. According to one study,[51] stereotype motions resulted in an increase in degenerative findings on radiographies. However, there was no evidence of disability or symptoms associated with these findings. Inflammatory osteoarthritis of the carpal bones is a common condition especially in middle aged females.[50] It may flare up after activities and appear to be work related, but this is probably only a temporary aggravating factor.

Irrespective of the etiology, the presence of arthritis can present substantial difficulty for those engaged in repetitive tasks. Also, it can cause confusion with other conditions such as inflammatory osteoarthritis, or rheumatoid arthritis of the carpal bones, or can result in carpal tunnel syndrome. Early recognition of this and treatment of the underlying disorder is, of course, important in understanding the condition.

GANGLIONIC CYSTS

Ganglionic cysts are very common and exist in a high proportion of the population. According to one study[17] there was no relationship in their prevalence to type of work. Ganglion cysts can be of several types; those that involve the synovial membranes of the wrist joints, or those that involve the tendon and tendon sheath. In the case of the latter, degeneration of the tendon and tendon sheath may result in swelling and cyst formation. It is biologically plausible that overuse may contribute to or cause some of these. These cysts, except for the pressure that they create, are usually painless. However, individuals often want them removed. Surgical removal in this case is the treatment, but they do tend to recur unless fully extirpated. Another treatment which may be helpful is injection with a needle, drawing out the fluid, and injecting a cortical steroid. However, these tend to recur fairly frequently.

REFLEX SYMPATHETIC DYSTROPHY

Reflex sympathetic dystrophy, sometimes called the "shoulder-hand syndrome", results from an overactivity of the sympathetic nervous system in the affected extremity. It may occur after injuries, especially in the shoulder region, or after a heart attack. Symptoms include burning type pain and swelling, with sensitivity to cold. CTDs, if not treated early and effectively, may progress to reflex sympathetic dystrophy. Reflex sympathetic dystrophy may occur after, or be aggravated by, carpal tunnel surgery. Treatment involves extensive physical therapy and sometimes stellate ganglion blocks. Calcium channel blockers such as nifedipine, and antidepressants such as amitriptyline may be helpful.

REFERENCES

1. Hippocrates, *The Genuine Works of Hippocrates*, translated by Francis Adams (London: Sydenham Society), Vol. 1. 1849:698.
2. Ramazzini, B., *The Disease of Workers*, (Chicago: University of Chicago Press),iIn W. Wright (trans.), 1717 and 1940.
3. Fry, J. H., "Overuse Syndrome in Musicians — 100 Years Ago, A Historical Review," In *The Medical Journal of Australia*, 145:620–625 (1986).
4. Haward, W., "Note on pianists' cramp," *Br. Med. J.*, 1:672 (1887).
5. Poore, G. V., "An analysis of 93 cases of writers' cramp and impaired writing power, making, with 75 cases previously reported, a total of 168 cases," *Br. Med. J.*, 1:935–936 (1887).
6. Down, J. L. H., "On the influence of the sewing machine on female health," *Br. Med. J.*, 1:26–27 (1866).
7. Rivers (for Hughlings Jackson), "Reports to the Medical Society of London," *Lancet*, 1:434 (1891).
8. Fry, J. H., "Overuse syndrome of the upper limb in musicians," *Med. J. Aust.*, 144:182–185 (1986).
9. Putz-Anderson, V., Ed., *Cumulative Trauma Disorders: A Manual For Musculoskeletal Diseases of the Upper Limbs.* (London: Taylor & Francis, 1988).
10. Miller, M. H. and Topliss, D. J., "Chronic upper limb pain syndrome (Repetitive Strain Injury) in the Australian workforce: a systematic cross-sectional rheumatology study of 229 patients," *J. Rheumatol.*, 15:11, 1988. pps. 1705–1711.
11. Silverstein, B. A., Fine, L. J., and Armstrong, T. J., "Hand wrist cumulative trauma disorders in industry," *Brit. J. Ind. Med.*, 43:779–784 (1986).
12. Peach, P. E., "Overwork weakness with evidence of muscle damage in a patient with residual paralysis from polio," *Arch. Phys. Med. Rehabil.*, 71:128–250 (1990).
13. Smythe, H. A., "Non-articular rheumatism and the fibrositis syndrome," in *Arthritis and Allied Conditions*, McCarty, Hollander, J. L. and D. J., Ed. (Philadelphia: Lea & Febiger, 1972) pp. 874–884.
14. Yunus, M., Masi, A. T., Calabro, J. J., Miller, K. A., and Feigenbaum, S. L., "Primary fibromyalgia (fibrositis): clinical study of 50 patients with matched normal controls," *Seminars and Rheumatism*, 11:151–171 (1981).
15. Dennett, X. and Fry, J. H., "Overuse syndrome:a muscle biopsy study," *Lancet*, 331:905–908 (1988).
16. Hadler, N. M., "Cumulative trauma disorders, an iatrogenic concept," *J. Occup. Med.*, 32(1):38–41 (1990).
17. McCormack, R. R., Inman, R. D., Wells, A., Bernste, C., and Imbus, H. R., "Prevalence of tendonitis and related disorders of the upper extremity in a manufacturing workforce.," *J. Rheumatol.*, 17:958–64 (1990).
18. Travell, J. G. and Simons, D. G., *Myofascial Pain and Dysfunction — The Trigger Point Manual.* (Baltimore: Williams & Wilkins, 1983)
19. Hurst, L. C. , Weissberg, D., and Carroll, R. E., "The relationship the double crush to carpal tunnel syndrome (an analysis of 1,000 cases of carpal tunnel syndrome)," *J. Hand Surg.*, 10B(2):202–204 (1985).

20. Eason, S. Y., Belsole, R. J., and Greene, T. L., "Carpal tunnel release:analysis of suboptimal results," *J. Hand Surg.*, 10B(3):365–369 (1985).
21. Roos, D. B., "Experience with first rib resection for thoracic outlet syndrome," *Ann. Surg.*, 173:429–442 (1971).
22. Swift, T. R. and Nichols, F. T., "The droopy shoulder syndrome," *Neurology*, 34:212–215 (1984).
23. Clein, L. J., "The droopy shoulder syndrome," *CMA Journal*, 114:343–344 (1976).
24. Lederman, R. J., "Thoracic outlet syndromes/review of the controversies and a report of 17 instrumental musicians," *Medical Problems of Performing Artists*, :87–91 (1987).
25. Adson, A. W. and Coffey, J. R., "Cervical rib: a method of anterior approach for relief of symptoms by division of the scalenus anticus," *Ann. Surg.*, 85:839–857 (1927).
26. Phalen, G. S., "The carpal-tunnel syndrome/clinical evaluation of 598 hands," *Clin. Orthop.*, 83:29–40 (1972).
27. Hadler, N. M., Ed., Bunn, W. B., and Co-ed., "Work-related disorders of the upper extremity Part II: Can shoulder periarthritis, thoracic outlet syndrome, or carpal tunnel syndrome be ascribed to repetitive usage?" *Occupational Problems in Medical Practice*, 4(3):1–8 (1989).
28. Armstrong, T. J., Ergonomics Guides/An Egonomics Guide to Carpal Tunnel Syndrome, American Industrial Hygiene Association, 1983.
29. Silverstein, B. A., Lawrence, J. F., and Armstrong, T. J., "Occupational Factors and Carpal Tunnel Syndrome," *Am. J. Ind. Med.*, 11:343–358 (1987).
30. Armstrong, T. J., Foulke, J. A., Joseph, B. S., and Goldstein, S. A., "Investigation of cumulative trauma disorders in a poultry processing plant," *Am. Ind. Hyg. Assoc. J.*, 43:103–116 (1982).
31. Morse, L. H., "Repetitive motion musculoskeletal problems in the microelectronics industry," *State of the Art Reviews: Occupational Medicine*, 1(1):167–175 (1986).
32. Feldman, R. G., Travers, P. H., Chirico-Post, J., and Keyserling, W. M., "Risk assessment in electronic assembly workers. carpal tunnel syndrome," *J. Hand Surg.*, 12A (2 Pt 2), 187, pps. 849–855.
33. Barnhart, S. and Rosenstock, L., "Carpal tunnel syndrome in grocery checkers," *West J. Med.* 147:37–40 (1987).
34. Margolis, W. and Kraus, J. F., "The prevalence of carpal tunnel syndrome symptoms in female supermarket checkers," *J. Occup. Med.*, 29(12):953–956 (1987).
35. Finkel, L. M., "The effects of repeated mechanical trauma in the meat industry," *Am. J. Ind. Med.*, 8:375–379 (1985).
36. Punnett, L., Robins, J. M., Wegman, D. H., and Keyserling, W. M., "Soft tissue disorders in the upper limbs of female garment workers," *Scand. J. Work Environ. Health*, 11:417–425 (1985).
37. Conner, D. E. and Kolisek, F. R., "Vibration-induced carpal tunnel syndrome," *Orthopaedic Review*, XV(7):49–54 (1986).
38. Nathan, P. A., Meadows, K. D., and Doyle, L. S., "Occupation as a risk factor for impaired sensory conduction of the median nerve at the carpal tunnel," *J. Hand Surg.*, 13B(2):167–170 (1988).

39. Gelberman, R. H., Rydevik, F. L., Pess, G. M., Szabo, R. M., and Lundborg, G., "Carpal tunnel syndrome — a scientific basis for clinical care," *Orthop. Clin. N. Am.*, 19(1,):115–124 (1988).

40. Stevens, J. C., "AAEE Minimonograph #26: The Electrodiagnosis of Carpal Tunnel Syndrome," *Muscle Nerv.*, 2:99–113 (1987).

41. Kimura, I. and Ayyar, D. R., "The carpal tunnel syndrome: electrophysiological aspects of 636 symptomatic extremities," *Electromyogr. Clin. Neurophysiol.*, 25:151–164 (1985).

42. Green, D. P., "Diagnostic and therapeutic value of carpal tunnel injection," *J. Hand Surg.*, 9A:850–854 (1984).

43. Ellis, J. M., et al. "Response of Vitamin B-6 deficiency and the carpal tunnel syndrome to pyridoxine," *Proc. Nat. Acad. Sci. USA*, 79:7494–7498 (1982).

44. Wolaniuk, A., Vadhanavikit, S., and Folkers, K., "Electromyographic data differentiate patients with the carpal tunnel syndrome when double blindly treated with pyridoxine and placebo," *Res. Commun. Chem. Pathol. Pharmacol.*, 41(3):501–511 (1983).

45. Folkers, K., Wolaniuk, A., and Vadhanavikit, S., "Enzymology of the response of the carpal tunnel syndrome to riboflavin and to combined riboflavin and pyridoxine," *Proc. Nat. Acad. Sci. USA*, 81:7076–7078 (1984).

46. Amadio, P. C., "Pyridoxine as an adjunct in the treatment of carpal tunnel syndrome," *J. Hand Surg.*, 10A:237–241 (1985).

47. Kasdan, M. L., and Janes, C., "Carpal tunnel syndrome and Vitamin B6," *Plast. Reconstr. Surg.*, 79(3):456–462 (1987).

48. Amadio, P. C., "Carpal tunnel syndrome, pyridoxine, and the workplace," *J. Hand Surg.*, 12A (2 Pt 2):875–880 (1987).

49. Parry, G. J. and Bredesen, D. E., "Sensory neuropathy with low-dose pyridoxine," *Neurology*, 35:1466–1468 (1985).

50. Ehrlich, G. E., "Inflammarotry osteoarthritis/the clinical syndrome," *J. Chron. Dis.*, 25:317–328 (1972).

51. Hadler, N. M., Gillings, D. B., Imbus, H. R., Levitin, P. M., Makuc, D., Utsinger, P. D., Yount, W. J., Slusser, D., and Moskovitz, N., "Hand structure and function in an industrial setting/influence of three patterns of stereotyped, repetitive usage," *Arthritis Rheum.*, 21(2):210–219 (1978).

ADDITIONAL READING

Cailliet, R., *Shoulder Pain,* Edition 2 (Philadelphia: F.A. Davis, 1981).

Management of Common Musculoskeletal Disorders — Physical Therapy Principles and Methods, (Philadelphia: Harper & Row, 1983).

Ryan, J. D., Ed. *Phyical Examination of the Musculoskeletal System,* (Year Book Medical Publishers, 1987).

Sheon, R. P., Moskowitz, R. W., and Goldberg, V. M., *Soft Tissue Rhematic Pain —Recognition, Management, Prevention,* (Philadelphia: Lea & Febiger, 1987).

Travell, J. G., and Simons, D. G., *Myofascial Pain and Dysfunction — The Trigger Point Manual,* (Baltimore: Williams & Wilkins, 1983).

PART II
SOLUTIONS TO A
MULTIFACETED PROBLEM

Chapter 4
The Basic Intervention Philosophy

CTDs comprise a significant amount of the workers compensation dollar and the problem is ever increasing. Recently, OSHA estimated that by the year 2000 over 50% of the workers compensation dollars will be as a result of Cumulative Trauma.

This statement should encourage companies to strive harder to *control* this problem, not only to contain costs but also to improve the health and welfare of their employees. But how is this done? CTDs are sometimes very difficult to evaluate. Several factors may lead to their inception and then other factors may contribute to their progression. Management of CTDs requires a multidisciplinary systems approach to control both the contributing and aggravating factors. The systems approach is driven by a broad base of knowledge, the ability to distribute responsibility effectively within the company, and a knowledgeable, cooperative medical provider system. The best results are achieved when: (1) preventative steps can be taken to lower the risk of the disorders, (2) there can be early recognition and treatment when problems do occur, and (3) a "rapid response" system re-addresses each individual situation to prevent progression and re-injury.

Who is responsible for this *control*? In a number of companies, this responsibility gravitates to the insurance carrier/third party administrator. In some, the physicians treating the problems may be given the responsibility. Sometimes the affected employee actually directs the care. **It is felt this responsibility should be accepted by the company itself**. Only in this way can a focused and responsible control be established.

The company is in the best position to address the elements necessary for control and thus, can more effectively reduce their lost time and workers compensation costs.

1. The company is in the best position to identify and modify risk.
2. The company is in the best position to recognize and treat the problems *early*.
3. The company is in the best position to enforce a rapid response system to re-address risk as inevitable problems surface.

If control is assumed, the plant can, in effect, close the revolving door that all too often creates disability from noncomplicated CTDs.

Insurance companies vary in their ability to help companies control CTDs. Few do a good job with *all* of the aspects outlined. The ability to handle the preventive, or ergonomic, aspect tends to vary with the experience of the "loss control" consultant. The capability to provide adequate early identification and treatment is often neglected when the insurance company alone is managing a CTD control program. An insurance company may not review the injured worker's job until the claims have been paid and the reserves have been sufficiently depleted to raise a "flag". In this instance, the worker may be allowed to progress to significant disability before the loss control function is effected.

It is unfair to expect the treating physician to oversee the quality of a company's CTD control program. Whereas most physicians try to do the best thing for the patient, most have very busy practices and just do not have the time to insure that one of their "client" companies is doing the job as well as it should. Even though a "plant physician" may occasionally visit the plant, it is very difficult for him/her to totally understand all the jobs and contact the appropriate supervisor for each problem that arises. The science related to CTDs is rapidly evolving. Many physicians are not aware of conservative, nonsurgical control measures that can be established *within* the plant.

For the above reasons, it is felt the *company* should accept responsibility for the lost time and cost associated with CTDs. Ultimately, the company has the most to gain and is in a position to be the most effective focus of *control*.

A SYSTEMS APPROACH TO MANAGING CTD

A systems approach involves a focused *control* at three stages in the development of CTDs.

1. Pre-event — primary prevention
2. Event — secondary prevention
3. Post event — tertiary prevention

Control is established by using two primary focal points — the supervisor and the medical department.

Pre-event — Primary Prevention

The most effective way to control an event is to remove or reduce the risk that allows the event to develop. This pre-event can be broken down into three components:

1. Identification of stressors
2. Implementation of controls
3. Monitoring of controls

A comprehensive review of objective, subjective, and historical information should be initially performed to prioritize the jobs in the plant according to the presence of ergonomic stressors. Beginning with the highest priority, the jobs should then be further studied to quantify posture, force, and recovery time stressors. Once quantified, these stressors aid in defining the "root cause" which forms the basis for establishing control mechanisms.

This process is best performed by an in-plant ergonomics committee which reflects a "team" of various perspectives in the workplace. Ergonomics is most successful when all levels and functions are involved in its study and control measure implementation. Close communication between hourly, supervisory, medical, engineering, and management functions are vital when trying to acquire an "ergonomics mentality" when implementing change. This committee should engage in the following activities:

1. Perform CTD risk prioritization and ergonomic job analyses
2. Monitor trends in the report of CTD cases, to aid in identifying causation
3. Ensure that "rapid response" audits of symptomatic employees are performed, and assist in any engineering/administrative controls that are indicated
4. Implement pilot projects and workplace and job modifications to address ergonomic stressors in the workplace
5. Document projects and activities to measure success

The ergonomics committee should attempt to transcend labor/management lines and simply function as an action team for problem identification and solution.

Event — Secondary Prevention

As long as physical work is performed, occasional CTDs may occur in spite of successful ergonomic intervention, short of mechanization. The objective of secondary intervention is to identify any problem as quickly as possible, in its earliest stage, to prevent it from becoming more severe. A system of early recognition, medical evaluation, and specific treatment must be established. Secondary prevention is easier to administer with an in-plant medical provider, such as a nurse or physical therapist.

Education is the most important strategy for successful secondary prevention. Education must involve all levels of the plant: upper management, middle management, workers, and technical support personnel. Nurses and/or physical therapists must be able to accurately evaluate and treat musculoskeletal complaints and physical findings. Upper management must be supportive of the people and time devoted to the control system. Middle management must oversee the program's implementation. Educating the worker improves body awareness, encourages active involvement in the ergonomic studies, and urges early reporting of fatigue or unusual symptoms.

The common educational theme should be that simple situations with easy solutions can become complex problems, if ignored. The main educational objectives are

1. To enlighten everyone in the plant on these simple, yet complex, medical conditions known as CTDs
2. To stress that focused early treatment can, in most instances, prevent surgery and minimize disability
3. That ergonomics must be involved in resolving stressors for the overall plant *and* each CTD occurrence.

Symptoms and complaints in the hand can result from physical problems arising anywhere from the neck to the tip of the fingers. In order to accurately evaluate the symptoms' etiology, the in-plant medical provider should address this entire possibility. "Tunnel vision" can tempt one to address only the hand and wrist area, but often leads to inaccurate diagnosis and has historically resulted in unsuccessful, and often multiple, surgeries.

Early and accurate problem identification allows effective, early treatment. Simple treatments such as anti-inflammatory medication, ice, massage, and stretching exercises, can resolve well over 50% of CTDs if they are recognized early, correctly assessed, and aggressively treated. *Adequate* treatment is better controlled if a significant portion is done in-plant, under the nurse's or physical therapist's supervision. Timely re-assessment of the condition must occur, with appropriate physician referral if no improvement is noted. An educated, coopera-tive medical referral system must also be established to guarantee a cooperative effort between physician, injured worker, and the plant.

Post Event — Tertiary Prevention

Risk reduction (primary prevention) accompanied by early problem recognition and treatment (secondary prevention) are essential in controlling the risk and the *costs* associated with CTDs. For the control of *incidence*, however, a system of checks and balances must be in place. There must be a mechanism to re-address risk for each individual that experiences a CTD. If such a mechanism is not in place, a "revolving door" effect may occur; the CTD symptoms may improve with treatment, but will eventually return if the cause has not been addressed.

A "rapid response" system coordinates the medical and prevention program components, and recruits the assistance of the front-line supervisors. In rapid response, the supervisor evaluates the ergonomic stressors affecting the symptomatic worker and develops an intervention strategy to minimize those stressors. The ergonomics committee can also assist in this response, especially if intervention requires workplace modifications. If the stressors can be addressed with retraining or a simple maintenance "work order", the supervisor is expected to initiate this activity immediately.

SUMMARY

The effect of CTDs on workers' compensation costs is significant today and experts predict that costs will only multiply if not addressed. It is imperative that companies establish a mechanism of control for this apparent "epidemic", utilizing a basic systems approach. The system involves a cooperative "team" of occupational physicians, ergonomists, nurses or in-plant medical care providers, "plant doctors", neurological and surgical consultants, and most importantly, trained in-plant personnel. Checks and balances must be put in place, and the system should be regularly audited for fine-tuning.

Without such a multidisciplinary systems approach, CTDs will continue to be handled in a "hit-or-miss" fashion.

Chapter 5
Ergonomics For Prevention

Ergonomics is the science of fitting the environment, work, and workplace to the physical capabilities and limitations of the working population. Ergonomics involves many disciplines, such as anthropometry, biomechanics, physiology, engineering, and industrial hygiene and safety. Ergonomic theory is rooted in all of these disciplines. Designing equipment and workstations requires research into anthropometric data reflecting population strengths and statures to determine what is acceptable to the widest range of workers. Establishing administrative controls will involve a good working knowledge of industrial engineering, training methodology, and exercise physiology.

This book is not intended to act as a handbook on the specifics of ergonomic controls. This chapter, instead, focuses on establishing a *systematic approach* to employing ergonomic controls in any given work situation and provides a brief review of the goals and intended results of various engineering and administrative controls.

IDENTIFYING THE ROOT CAUSE

It is important to avoid a "quick and dirty" dash through the plant to identify various postures, forces, and repetitive situations. These situations will exist in any work setting! The band-aid "fix it" approach must be discarded and replaced with a systematic review of the available information to determine the existence of ergonomic stressors. In other words, the root cause must be identified before controls can be implemented.

Three primary perspectives exist to uncover ergonomic stressors and their root causes in a work setting. They are

- Objective perspective: audit of job components
- Historical perspective: existing data on past and present CTDs
- Subjective perspective: opinion of the people who perform the job

PRIORITIZING JOBS

Prior to an in-depth study, time can be saved if each job is briefly audited to identify which jobs have the most ergonomic stressors impacting the most body parts in the most severe combinations. Health & Hygiene, Inc. uses a simple scaled ranking system for each body part, ranking from 0 (no force, posture, or repetition concerns) to 5 (extreme postures, heavy loads, minimal recovery time) to determine a level of "risk". In this way, jobs are prioritized by *anticipated*"risk" in the greatest number of body parts or in a specific body part. This initial scoring system effectively teases out higher from lower risk jobs, and aids in a systematic prioritization of jobs for further study. Not surprisingly, many jobs score with "high" risk potential but have yet to result in any injuries, illnesses, or worker complaints. If only subjective and/or incidence rate data had been consulted, then those potentially high risk jobs would have gone undetected. Such a system creates an additional information source to identify *potentially* stressful jobs.

Next, gather information reflecting the historical perspective. Review the OSHA 200 logs, nurse data sheets, first-aid logs, and other medical records to identify CTDs that have previously occurred. Accurate record keeping is essential to obtain a complete picture of the body part affected, the severity of the injury, and the specific job associated with the CTD. Logs are useless if the job is described as "general laborer", without further delineating a specific job title or task. On the logs, other tell-tale symptoms may not directly reflect a CTD diagnosis, but will hint of muscle fatigue: headaches, repeated requests for aspirin, and an unusually high incidence of cuts and contusions (indicative of increased clumsiness from fatigue). Other significant "symptoms" include jobs that have chronically high turnover rates, absenteeism problems, or high error/rework percentages.

Jobs associated with injuries, illnesses, and suspicious trends should then be prioritized by number of lost-time workdays. However, do not discount occurrences that result in job transfer or restriction, even if no work days were lost. Many plants respond to a worker's complaint of fatigue or pain with a job transfer, which may simply mask the root problem.

Finally, collect the "expert" perspective — that of the worker who must actually perform the job. Workers provide valuable information about the job's difficulty and usually can recommend reasonable solutions. Prior to any ergonomic audit or study, workers should be notified of its intention so their cooperation and assistance can be obtained. Interview techniques, pain diagrams, paired comparison and exertion scaling questionnaires all are good tools to objectively reflect the worker's subjective feelings about a job.[1-3] Subjective information should never be used as the single means of identifying hazardous conditions in a work environment. Attempt to objectify subjective information gleaned from interviews and group meetings with numerical scaling techniques. Workers have been known to manipulate subjective information to enhance union campaigns, wage negotiations, and other areas of labor dispute.

All three perspectives bring fresh information about problematic jobs. Each act as a check and balance for the other two. It is not sufficient to assess ergonomic risk, identify root causes, and apply controls by simply using one perspective. If one performs only job audits, then stressful jobs may be overlooked because "the workers make it look so easy". Medical data and historical information may not reflect the true stress of a job due to natural worker migration to lighter or higher paying jobs, or if medical personnel are readily accessible. An active medical program creates higher usage and more entries on the OSHA 200 and first-aid logs. Therefore, a plant without aggressive on-site medical resources may not realize that muscular fatigue and cumulative trauma exist. Either the workers do not report them, or they seek care outside of the employer's jurisdiction.

In summary, one must use a systematic approach to accurately evaluate CTDs. Spend an adequate amount of time to uncover trends, then merge different data sources to establish a priority system for further job review. Obviously, each job in a plant cannot be studied or redesigned at the same time. The sensible approach is to attack the worst offenders first. Good communication with workers is essential to maintain a positive atmosphere, so that workers performing jobs lower on the priority list do not become impatient.

IN-DEPTH ANALYSIS

Beginning with the highest priority, the jobs should then be individually studied to quantify posture, force, and recovery time stressors. Videotaping is extremely useful to perform a good posture analysis, as most video cassette recorders now have slow motion capabilities. If video cameras and players are not available, then simple industrial engineering work sampling methods will suffice to identify extreme postures. Time and motion studies and productivity information are vital for identifying work/rest ratios. Exertion ranking scales, pinch and grip dynamometers, scales, and equipment manufacturer information will all serve to reasonably estimate force requirements.[3,4]

Tape measures and blue prints are necessary for obtaining layout and dimensional information. Key dimensions to include are: work height, reach distances, vertical and horizontal carrying distances in manual material handling tasks, walking distances, foot and knee clearances under tables, and tool, product, and employee (elbow height, stature, eye level) dimensions.

Once the stressors (force, posture, and recovery time issues) have been quantified, the significance of those stressors can be determined. If significant, a root cause must be identified in order to effectively control the stressor. It may be helpful to review the summary list of root cause sources discussed in Part I, Chapter 2. Many root causes may be evident. It is important to associate the cause to the stressor identified. Although footrails are always a nice addition for standing jobs, they may do nothing to reduce the impact of wrist deviation or poorly designed tools on the hand and forearm.

ERGONOMIC CONTROLS

Two types of controls are generally available to remedy a workplace stressor. *Engineering controls* are utilized to modify the workplace, tool, or job design so that the posture, force, and recovery time stressors are minimized or eliminated. *Administrative controls* are utilized to minimize exposure to the known stressor when engineering controls are either impossible or economically infeasible. Always consider engineering controls initially; they have the greater impact on modifying a worker's behavior and performance in the workplace. Administrative controls require constant administrative monitoring to ensure compliance from both management and workers. Many times, administrative controls are used as interim controls while engineering controls are planned and implemented. This is acceptable and reflects responsible action from the employer, as the administrative control deals with a known stressful condition until it can be permanently addressed. Workers often have difficulty with the time expanse between efforts to prioritize, study, and identify stressors, and the implementation of control measures. Consistent and continuous communication during the planning and implementation of engineering controls will ensure that workers do not feel neglected or ignored. Regular updates on a project's progress and requests for worker input minimize unnecessary disputes during this very beneficial and proactive activity.

ENGINEERING CONTROLS

Many reference texts are available for specific assistance in anthropometry, tool specifications, and engineering methods. From these texts, basic guidelines are listed below which can be followed with good results and should act as minimum requirements for any workplace design.[5-12]

Guidelines for Workstation Design

1. Orient the workstation so the operator faces the work, without twisting, turning, or bending.
2. The worksurface height should accommodate work type and visual requirements.

Work type	In reference to elbow height
Delicate/fine work, writing, assembly	2–4 in. above
General handwork	2–4 in. below
Manual work, using tools	4–6 in. below
Heavy manual work, using upper back and legs	6–16 in. below

Additionally, angle the worksurface if visual acuity must be further enhanced.

3. Rule of thumb — build in adjustability. If unable to adjust, then design according to population percentiles:

    *   Establish worksurface height for 50th to 95th percentile worker. Raise shorter workers with workstands.
    *   Establish reaches and shelf/conveyor heights for 5th percentile female (smallest worker).

4. Orient all incoming/outgoing product containers for easy access in (angle) and to (height).
5. All aside movements should be down and out, within 14 to 18 in. reach, at working height or slightly below.
6. Design area so movement occurs in the same horizontal plane — all objects to be manipulated should be arranged so that the most frequent movements occur with the *elbows bent and close to the side.*
7. For optimal hand/arm strength and skill, work should be placed 10 to 12 in. in front of eyes, and elbows should be postured at an 85 to 100° angle.
8. If object needs to be higher and closer for visual acuity, use supports under the elbows, forearms, hands. These supports should be padded and adjustable.

Guidelines for Handtool Design

1. Handles must keep the upper extremity joints in neutral positions. Avoid deviated wrists and elevated elbows.
2. If the center of gravity is located to one side or forward (e.g., motor, gear shaft), use two handles for balance.

3. The center of gravity of the tool should be located as close to the body as possible (improves ease in handling).
4. If long and wide, use two handles for ease of control.
5. If the tool weighs over 5 lb and will be used for a majority of the shift, use two handles.
6. Heavier tools (> 25 lb) should have two handles, be suspended, and counter balanced.
7. The handle length should accommodate the 95th percentile hand breadth (4 to 5 in.).
8. Handle diameter should be at least 1.5 in.
9. Two-handled grip span should equal 2.6 to 3.5 in., as this is the grip distance that maximizes grip strength capability.
10. Handles should provide good electrical and heat insulation. Compressible rubber and plastic are good materials for this purpose, as well as for hygiene purposes.
11. Handles should not have a smooth, polished surface. A longitudinal grooving or texturing to the surface allows a good, nonslippery contact.

Guidelines for Controls and Displays

1. Controls and displays should be easy to see, reach, and grasp with *neutral* neck and arm postures.
2. Controls and displays should be grouped functionally.
3. Displays should be close to the control: over, above, or to the *right* of the control.
4. If controls and displays must be in two separate panels, then they should be arranged exactly the same way.
5. ID labels should be placed *above* the control and *above* the display.
6. If controls are operated in a sequence, arrange them in that order from *left to right*.
7. Use standardized, human stereotyped patterns, color, language. Numbers should progress from left to right — top to bottom.
8. If a control requires a quick response or large effort, provide a large surface. It should be big enough to be activated with confidence.
9. Choose foot pedals over knee controls, but foot controls should be used when seated or in a sit/lean position.
10. When standing, knee or hand controls are preferred to foot controls.

Guidelines for Seated Work

1. Workstation height should be adjustable (by lowering or raising the height of the worksurface or the height of the worker's chair). Footrests should be provided, as needed.
2. Characteristics of a good chair:

- No tools are needed to make adjustments
- Lumbar support
- Arm rests (if used) placed well back from front of seat pan
- Pneumatic height adjustment
- Five legs for stability
- Firm cushions covered in a nonslip fabric
- Rounded "waterfall" front

3. The chair should be adjusted so that:

   - The body's weight is evenly distributed on the hips and thighs
   - The front of the seat pan does not press against the backs of the knees or lower legs
   - Work surface or home row on the keyboard is at elbow height

4. Characteristics of a good desk:

   - Leg movement is not obstructed
   - The working surface allows sufficient room for the operator's work, including electronic equipment and work aids
   - Work surface heights may vary from one part of the desk to other depending on the task performed and equipment used (i.e., keyboard at elbow height, writing surface at midchest height)
   - Provides sufficient horizontal and vertical storage

5. Furniture and equipment should be arranged to promote an uninterrupted flow of material without constant twisting, bending, or getting up.

   - Files and binders used most often should be at fingertip access.
   - Infrequently used items can be stored farther away.

6. Employees working at VDTs should occasionally look away from the screen at an object that is at least 6 ft (2 m) away.
7. Place the VDT screen at a distance of about 20 in. (50 cm) from the operator's eyes.
8. Height and screen angle of the VDT should be adjustable.
9. Use copy holders to position documents. If keying from hard copy, place the holder in front of the operator. If editing from the screen, the holder should be placed to the side.
10. The angle and height of the keyboard should be adjustable.
11. Avoid glare:

    - Do not face a window directly or place a window behind a VDT screen.

- Place desks perpendicular to windows.
- Ceiling light fixtures should be located to the side of the work surface, not directly over or in front of it.
- Remove bulbs or fluorescent tubes that shine directly into the eyes or cause direct glare from the work surface.
- Use task lights to further illuminate the work surface.

12. Do not sit near excessively cold or hot surfaces and avoid drafts (especially at neck or feet).
13. Office temperatures should be about 68 to 72°F (20 to 22°C) in the winter and 72 to 75°F (22 to 24°C) in the summer.

ADMINISTRATIVE CONTROLS

Administrative controls can be effective if implemented properly. If implemented without a thorough understanding of rationale and effect, supervisors will have difficulty devoting their time to oversee administrative controls, and workers will not regard the controls seriously. Five primary forms of administrative controls can be discussed:

1. Establishing proper work methods
2. Training employees
3. Work pacing and scheduling
4. Exercise programs
5. Job rotation and enlargement

Establishing Proper Work Methods

Ergonomically acceptable work methods should be established for each job by obtaining input from both supervisors and employees. Videotaping efficient and inefficient workers helps to identify different styles and necessary and unnecessary motions. Industrial engineering time and motion studies can then determine the most efficient method. Hand position, handling technique, carrying and reach distances, maintenance procedures for ease of equipment and tool use, and quality issues should also be identified.

A pictorial booklet or instructional video displaying correct methods, incorrect methods, and common mistakes should then be developed for future reference by managers, trainers, and workers.

Training New Employees

It is important to invest in a formal training program for new employees. A Swedish study showed that the number of days lost through neck, shoulder, and arm complaints was reduced by 50% in new workers participating in an ergonomic work methods program. Experienced workers participating in the same program showed no significant improvement.[13] Ergonomic methods instruction must be done when new workers are learning the motor patterns required of their new tasks.

A formal training program should include a training supervisor or manager, and preferably on-line trainers with the same status as bid jobs, lead employees, or assistant supervisors. Hiring criteria is crucial for these trainers and should be pre-established, especially if given a bid job status. Utilize trainers to work with new hires, transferees, employees with poor job methods, CTD symptomatic employees, and employees returning from vacation or medical leave. They can also oversee cross training efforts and assist supervisors in managing job rotation or other administrative control efforts. Trainers should *not* be considered as fill-ins or utility employees. Standard methods should be known and used by all trainers so as to exhibit a good example.

All information on proper work methods should be translated to new employees, along with any other safety or orientation information, *prior* to actually performing the job. Visual feedback using videotape examples of good techniques and common mistakes will help to cue the new worker for future performance reference. Trainers should establish documentation regarding trainee instruction, proficiency, and progress throughout the training period.

New employees should then work off-line with the trainers to learn the skills of the job. Once proficient with the skills, place the worker on-line to develop speed, preferably with an experienced employee to gradually condition into full speed. Part II, Chapter 7, provides further information about job conditioning. Depending on the complexity of the job, up to 4 to 6 weeks may be necessary for the employees to obtain full speed. In some complex sewing operations, for example, this training period extends into several months. When retraining experienced employees, apply the same procedure. However, 4 to 6 weeks may not be necessary unless a worker has been absent for an extended period of time, or is learning a completely new job.

A system should be established to document an employee's lack of compliance with standard methods, unsafe acts, or inability to perform the job after an adequate training period is exceeded. Check performance proficiency at regular intervals, and progress the trainee onto the next skill or speed level contingent upon capability to meet standard method, quality, and safety standards without undue muscle soreness.

These guidelines will provide structure and consistency to the training efforts. Benefits of on-line trainers include

- Supervisor freedom from time-intensive on-line training
- Consistency in training and managing new employees
- Reduced early turnover associated with production stress, soreness from inadequate conditioning, and inadequate feedback
- Higher skill and conditioning levels
- Improved quality and yields
- Provides a system to accomplish cross-training, which assists in job rotation plans and skill depth within a plant or department

Work Pacing and Scheduling

If work days are expanded to 12 h on a regular or overtime basis, modifications to the pace and schedule may be necessary to avoid fatigue. Jobs that require 33% energy requirements of a worker's aerobic capacity over an 8-h day would need to be modified so that the requirements are reduced to 28% to accommodate a 12-h day. If pacing is *just* acceptable for an 8-h day ("tight" standards), hourly production expectations should be decreased for the additional hours. Some references estimate that, based on production loss studies, 2.5% production decreases per additional hour should be allowed.[7]

If the existing pace is fully acceptable for an 8-h day ("loose" standards), then the pace may not require modification when moving toward extra hours. When considering overtime, a time and motion study can determine the effect of additional hours at the present work pace on the workers and is preferred to the typical trial and error approach.

If overtime leads to 6- and 7-d work weeks, with 1 d off every 2 to 4 weeks, the hiring of relief, evening, or swing shifts may be necessary. The economic feasibility of adding part-time swing crews and the impact on product quality should be compared to the safety and health costs and overtime pay.

Exercise Programs

Although exercise in industry is an old concept, exercise was recently promoted by Ethicon, Johnson & Johnson, as a prophylactic means for reducing CTDs. Both flexibility and strengthening exercises were performed on a set schedule throughout the day. Although published exercise program success stories do exist, two studies have found no prophylactic effect and dubious economic efficacy of exercise programs.[14–16]

However, these studies may naively discount exercise as an administrative control for cumulative trauma. The researchers investigated the *prophylactic* effect of exercise on the development of carpal tunnel syndrome, the end result of cumulative and prolonged muscle fatigue and overuse. The use of stretching exercise alone as a means to control fatigue and muscle soreness was not investigated.

The programs studied utilized both strengthening and stretching activities. Strengthening exercises are not appropriate activities to include during the work shift, as the workers are using physical strength and energy to perform their jobs. Adding strengthening exercises to a currently full workday robs a worker of necessary recovery time and only serves to use more strength and energy.

Alternately, stretching exercises are designed to maintain muscular mobility and circulation. By following sports examples, flexibility stretching is regularly used to aid performance and reduce the risk of injury. Stretching a fatigued muscle offsets cramping and accelerated fatigue by allowing the muscle to move in a different direction and infusing circulation and recovery time into the workday.

Employees should be encouraged to stretch muscles intermittently throughout the day when feeling fatigued or cramped, rather than allowing these sensations to worsen and develop into a CTD.

Job Rotation and Enlargement

Job rotation is considered an administrative control to limit employees' overall exposure to stressful or fatiguing jobs and is particularly useful for jobs requiring concentrated hand work. Job rotation is used for: (1) reducing psychological fatigue (boredom), (2) reducing physiological fatigue (musculoskeletal tension), and (3) limiting exposure to specific environmental stressors.[17]

If used as a preventative measure, job rotation must be administered generally and consistently, and *not* as an injury/illness-specific response for individual employees. Job rotation is *not* light or alternative duty work, but rather a means to structure the tasks in an overall workday. A form of rotation, *job enlargement*, combines several tasks into one cycle to add variety and a mixture of high and low stress activities to the job.

Job rotation should be established after carefully studying the jobs to determine: muscle groups and motions utilized, tools, positioning (sit, stand, etc.), exertion level (light, moderate, heavy), number of fundamental elements, and repetitiveness.

Limitations on employee travel, product batching, and materials movement should be established to determine rotation time frames.

Rotation should occur *within* the workday. Suggested rotation frequencies are

- 15 to 30 min for highly concentrated, repetitive jobs
- 30 to 60 min for moderately concentrated repetitive jobs
- At breaks or every 2 h for jobs where employees travel or product movement constraints exist within the plant

Pair jobs that are significantly dissimilar. Allow for adequate cross training of all participating employees.

The use of job rotation as an administrative control has created quite a controversy among ergonomic experts, labor union officials, and production management. Rotations are time consuming to initiate due to cross training efforts. Job rotation is easiest to administer in low-skilled jobs. Job rotation is most effective and easily applied in concentrated, single-task assembly processes, where workers are in close proximity to one another, and general knowledge of the total assembly process exists. The National Institute for Occupational Safety and Health, although frowning upon indiscriminate use of job rotation as an ergonomic control, has recommended the use of job rotation when other controls are not readily apparent or feasible.[18]

REFERENCES

1. Corlett, E. N. and Bishop, R. P., "A technique for assessing postural discomfort," *Ergonomics*, 19:175–182 (1976).
2. Putz-Anderson, V., *Cumulative Trauma Disorders: A Manual for Musculoskeletal Diseases of the Upper Extremity*. (New York: Taylor and Frances, 1988)
3. Wilson, J. R. and Corlett, E. N., Eds. *Evaluation of Human Work*, (Philadelphia: Taylor and Frances, 1990) Chapter 3.
4. Drury, C. G., "Biomechanical evaluation of the repetitive motion injury potential of industrial jobs," *Seminars in Occup. Med.*, 2(1):41–49 (1987).
5. ACGIH. *Ergonomic Interventions to Prevent Musculoskeletal Injuries in Industry*. (Chelsea, MI: Lewis Publishers, 1987)
6. Eastman Kodak Company, Human Factors Section. *Ergonomic Design for People at Work*, Volume I. (New York: Van Nostrand Reinhold Company, 1983)
7. Eastman Kodak Company, Human Factors Section. *Ergonomic Design for People at Work*, Volume II. (New York: Van Nostrand Reinhold Company, 1986)
8. Grandjean, E., *Fitting the Task to the Man*. (Philadelphia: Taylor and Francis, 1985)
9. Himmelstein, J. S. and Pransky, G. S., Eds. "State of the Art Reviews in Occupational Medicine," *Worker Fitness and Risk Evaluations*, 3:2. (Philadelphia: Hanley and Belfus, 1988)
10. Human Factors Society. *American National Standard for Human Factors Engineering of Visual Display Terminal Workstations*. (Santa Monica: The Human Factors Society, 1988)
11. Van Cott, H. P. and Kinkade, R. G., *Human Engineering Guide to Equipment Design*. (Washington, DC: U.S. Government Printing Office, 1972)
12. Woodson, W. E., *Human Factors Design Handbook*. (New York: McGraw-Hill Book Company, 1981)
13. Parenmark, G., Engvall, B., and Malmkvist, A-K., "Ergonomics on the job — training of assembly workers," *Appl. Ergon.*, June: 143–146 (1988).
14. Sawyer, K., "An on-site exercise program to prevent carpal tunnel syndrome," *Prof. Safety*, May: 17–20 (1987).
15. Silverstein, B. A., Armstrong, T. J., Longmate, A., and Woody, D., "Can in-plant exercise control musculoskeletal symptoms?" *J. Occup. Med.*, 3(12):922–927 (1988).
16. Williams, T. L., Smith, L. A., and Herrick, R. P., "Exercise as a prophylactic device against carpal tunnel syndrome," in *Proceedings of the Human Factors Society 33rd Annual Meeting*, (Santa Monica: The Human Factors Society, 1989) pp. 723-727.
17. Konz, S., *Work Design Industrial Ergonomics* (Columbus:Publishing Horizons, 1987) pp. 341–342.
18. "Job Rotation Suggested at Plant for Work Causing Highest Rate of Injury," *Occupational Safety and Health Reporter*, The Bureau of National Affairs, Inc. (1990), p.900.

Chapter 6
Effective Occupational Medicine
for Control

A serious complicating factor in the management of CTDs has been the lack of early recognition and early effective care. It is not uncommon when visiting plants to find employees who have had pain and other symptoms such as numbness for months or years. In recent years, the problem of lack of early recognition has been changing due to media publicity and employee education programs. However, early recognition alone is not enough. If early recognition results in early care by practitioners who are inexperienced in handling these problems, or by those who advocate surgery as an important means of care, it will only add an iatrogenic (physician-induced) complication to the problem. Early recognition without a sincere and informed attempt by management to deal with the work-related problem will only result in frustration of the care providers and hostility of the employee.

As already discussed, muscles become fatigued from overuse. Immediate or delayed soreness results. Delayed soreness results in spasm, and if healing is not allowed to occur, a pain-spasm cycle ensues. Myofacial pain in time may develop into a chronic fibromyalgia which is very resistant to treatment. Recognition and intervention during the early stages can resolve the problem in a high percentage of cases (80 to 90%). However, recurrences are prone to occur if attention is not directed to the biomechanical and ergonomic aspects of the work.

However, painful spastic muscles can lead to or complicate tendon injuries. Tendon injuries may occur as a result of continued biomechanical stress. Tendon injuries are much more resistant to treatment, and if inflammation is allowed to continue, can result in scarring and narrowing of the tendon sheath. Nerve entrapment syndromes if untreated, will frequently progress and can cause wasting of the muscles and long-term or permanent disability.

The above are bonafide physical reasons for early intervention. Also of great importance are psychological reasons. Workers should not have to work with chronic pain. The truth of this statement on humanitarian grounds alone is obvious. However, pain results in dysfunction, lack of productivity, absenteeism, and therefore is not only costly to the employee, but also to the employer.

Fear becomes an important component of these disorders. Employees see their co-workers who have become disabled from their work and have gotten into an adversarial workers' compensation system. Managers who do not have to use their muscles in order to make a living often have difficulty in understanding the fear that is created sometimes even by minor disorders. They, therefore, may believe that the employee is exaggerating and making a great deal of little. Consider, however, if you must use your right arm day after day to make a living. You may have a family to support or perhaps a sick spouse. Your arm becomes painful; numbness and other symptoms arise. You naturally ask "Am I going to become disabled? Am I going to end up like that person next to me who had to have surgery and then developed uncontrollable swelling every time he/she tried to use it? Will I lose my job? Will I be thrust into the adversity of the workers' compensation system?"

Depending upon the response of management, and the make-up of the employee, fear could give way to hostility. The employee thinks: "I have worked very hard for them. Now they doubt that I am telling them the truth about my pain and want to discard me." Legal advice is sought. Once the hostility and adversarial relationship becomes entrenched, the employee will often lose motivation to continue work. The result is prolonged disability.

Employers are often fearful that encouragement of early reporting and treatment will result in symptom exaggeration and malingering. Managers not infrequently comment that if employees are encouraged to report these problems early, the dispensary will be filled with "gold bricks" and malingerers. Although the minority of the employees who are in this category will come forth as a result of early recognition programs, they are soon recognized and will be dealt with appropriately by qualified medical personnel. It is far better to have even poorly motivated employees to receive early attention in the medical facility than to go outside on their own, into a system that often encourages disability.

It is beyond the scope of this book to discuss workers' compensation issues in any detail. In a number of states the workers' compensation system makes it very difficult to manage these cases. Employees may go to practitioners who are thoroughly unfamiliar with and unable to properly manage these problems. Legal impediments to prompt evaluation and treatment sometimes exist. Tax-free workers' compensation benefits are sometimes at a level where an employee earns the same income when not working. This may provide great temptation for the employee to stay home and collect compensation, especially where there may be small children or a sick spouse. However, early recognition and prompt effective care and management of cases will reduce lost time and disability for the majority of employees affected with CTDs.

APPROACHES TO EARLY RECOGNITION

Unfortunately at the present time, we do not have proven objective screening methods to provide early recognition of CTDs. Workers exposed to dust on the job can undergo pulmonary function testing, which at the least provides indication that lung dysfunction is occurring. Workers exposed to lead can have blood lead and other tests which indicate abnormal accumulation of the substance. For CTDs at present, screening methods revolve around symptom surveys, screening examinations, and screening tests.

Symptom Surveys

Symptom surveys have been advocated[1] and may result in a high proportion of employees reporting symptoms.[2] These surveys are accomplished by utilizing questionnaires which ask if the employee is experiencing pain or other symptoms. Ideally, they provide a means, either by question or diagram, for the employee to localize the symptoms. Other questions indicate the length of duration of the symptoms. It has been advocated that symptom surveys be done as an initial evaluation and at periodic intervals thereafter. These surveys can help to determine jobs which are most troublesome and provide guidance for ergonomic intervention.

Some advocate that screening surveys be anonymous, for the reason that employees will answer more truthfully if there is no fear of recrimination. On the other hand, anonymity makes it impossible for the employer to deal effectively with any individual employee's problems. In the study cited above,[2] 47% of employees reported that they were currently having some problem. This was in spite of highly charged and emotional atmosphere in which CTDs were a point of contention. Employees were not screened anonymously in this study.

A problem with screening surveys is that since they are so highly subjective, the percentage of positive results may at times be largely a reflection of a company's employee relations. Then when effective ergonomic programs are started, it is sometimes difficult to determine whether the improvement is due to the "Hawthorne" effect or through ergonomic improvement.

In spite of the above constraints, symptom surveys may provide baseline and postintervention data which will help to document the results of the program.

Screening Examinations

Physical findings in CTDs are often sparse. However, the experienced examiner may often detect early signs such as, trigger points, swellings, muscle spasm, and tenderness, limitation of motion of the joints, ganglion cysts, or calcification. It is doubtful that screening examinations will detect many conditions of which the employee was not already aware. Screening examinations can be done by a specially trained nurse. They are relatively time consuming. Therefore, these examinations are most appropriately used in conjunction with reports of symptoms or symptom surveys.

Screening Tests

There is no single objective test to tell whether a person is developing a CTD. Currently available tests include strength testing, nerve conduction testing, and quantitative sensory testing.[3]

Strength Testing

Strength can be a rough indicator of the function of the hand and forearm. It may be especially helpful in serial testing where one can repeatedly test the same individual over a period of time. Grip strength testing, though nonspecific, may be reduced in any condition that affects the strength of the muscles of the hand or forearm, specifically carpal tunnel syndrome, forearm myalgia, and fibromyalgia. Normally the grip strength of the dominant hand is about 15% greater than the nondominant. Therefore, in a right-handed individual, grip strength in the left hand that is greater than that of the right hand, is an abnormal finding. Individual values vary considerably, however, a set of normal values has been developed.[4] Pinch strength testing is also helpful. It is reduced if there is thenar muscle weakness in the carpal tunnel. There is considerable importance in how the pinch strength test is done. The pinch of the tip of the thumb to the index or ring finger should be tested.

Tests of Nerve Conduction

Nerve conduction tests at the median and ulnar nerve at the wrist are used in diagnosing carpal tunnel syndrome. Recently devices for in-plant screening of motor nerve conduction have become available. However, according to study,[5] over 40% of those with carpal tunnel syndrome have normal distal motor latency.[6] No study, to date, has been published of serial nerve conduction velocity tests over a period of time and their value in detecting early CTDs. Sensory nerve conduction can also be screened. It is more sensitive than motor testing, but may be subject to more variation. Some screening studies have also shown that a relatively high percentage of asymptomatic persons without evidence of carpal tunnel syndrome have abnormal latencies.[7]

Quantitative Sensory Testing

These tests[8,9] show the ability of the person to discern vibration at various frequencies and intensities. Vibration sensation may be impaired by a variety of lesions, including neuropathy, and entrapments at various levels of the upper extremity. Localization of the abnormality to the median nerve distribution of the hand is indicative of median nerve dysfunction, but does not localize the lesion. It also depends upon subject cooperation in indicating when the stimulus is felt.

Other Tests of Sensation

Instruments are available to test for the individual's ability to perceive electrical stimuli.[10–12] These tests, similar again, are not specific and principles as applied to the vibrometry testing apply here.

Thermal Sensitivity

Another instrument tests sensitivity to heat and cold sensation.[3,13] The discussion above is applicable.

Other Tests

There are a variety of other tests[14] that may be utilized including a simple two-point discrimination in which the observer measures the individual's ability to distinguish two vs one point at varying differences. The Semmes Weinstein monofilaments[15] standardize pressure as a means of determining sensibility.

All of the testing outlined above, including strength testing, have potential inaccuracies. Workers are brought in off a production line when working. They may have used upper extremity motions which increased the pressure in the wrist or decreased the circulation and thereby temporarily affect the test. They may have been exposed to cold which can reduce nerve conduction velocities. They may be fatigued with decreased strength. In order to avoid these potential difficulties, such testing, is best done before the beginning of a shift or after a period of rest. Research is needed to evaluate the effectiveness of various testing protocols.

All of the screening procedures, including questionnaire, examination, and testing have the difficulty that they can only be done periodically. Though they provide an opportunity for serial determinations, the employee may develop symptoms and problems in between the testing. They may be deluded into a false sense of security by a previous normal testing.

EARLY RECOGNITION AND REPORTING OF SYMPTOMS

Early recognition and the reporting of symptoms by employees is an effective means of surveillance. In fact, there is no data available that indicates screening by physical examinations or tests can detect problems before an informed employee will report them. Of course, any company that institutes an early reporting program must have a plan from beginning to end in how to manage that case. There must be effective, early in-plant and out-of-plant treatment and good case management. With these in place, promoting early reporting can provide great benefits to both the employees and the company. It shows employees that the company cares. The fact that the employee can report a problem without fear of recrimination or loss of job can defuse a tense situation which often results when there are CTDs in a plant.

In order to have an effective early reporting program the company must have a policy where employees are not penalized for reporting and this entails an active education program for the employees. Employees should be taught to distinguish the soreness that results from new activity, which disappears after one gets used to the effort, from soreness that persists. Generally, we like to tell employees that the following are abnormal and should be reported:

1. Soreness that is persisting after a couple of weeks on a new job.
2. Soreness that develops without explanation in an employee who has been on the same job for some time.
3. Soreness that is intensifying rather than improving.
4. Soreness that is localizing in a specific muscle.
5. Soreness that is accompanied by findings such as crepitation, spasm, or swelling.
6. Any signs of numbness or tingling.
7. Any redness, swelling, inflammation, or crepitation.

In-Plant Care

In-plant care is an important adjunct to the early recognition of CTDs. In-plant care is easy for the employee to use. It can be used promptly without waiting for doctors appointments. It can be done on company time, therefore, there is little resistance to its use. It is provided by someone who is familiar with the job. It avoids the employee unnecessarily getting into a medical care system which may not be sympathetic with or understanding of the problem. The downside that often worries management is that the employees will be coming to the evaluation and treatment facility in order to spend time away from work. With an experienced nurse or therapist, this may happen a few times with an employee, but generally will not continue. The other problem with in-plant treatment is that many companies do not have the size or the budget for a full-time nurse or therapist. Adequate facilities in many plants are often lacking.

Strategies for In-Plant Care

Companies that have a full-time nurse have a ready made foundation upon which to build an in-plant early care program. Plants that have 250 or more employees and have a considerable number of cases, or high workers' compensation costs from CTDs should consider a full-time nurse. This will likely be a very cost effective approach dealing with these problems.

Companies without a nurse should develop alternate strategies. Many hospitals throughout the country have occupational programs which will provide a part-time nurse. Some physicians' offices will contract for part-time nursing service. Some companies utilize a part-time physical therapist to develop a program.

Space must also be provided for in-plant evaluation and treatment. Unfortunately, many companies do not have this. Facilities need not be fancy — preferably two rooms should be available; one for examinations and another for treatments. Larger operations should have larger medical facilities in proportion to size. A rule of thumb is 1 ft^2 per employee with a minimum of 240 to 300 ft^2.[16,17]

The Role of the Occupational Health Nurse

The occupational health nurse can be a tremendously valuable asset in dealing with CTD problems. If the nurse is empathetic to the employees, they will report problems when they are encouraged. The nurse must not only be caring, but competent and business-like, and able to keep good records.

The nurse needs to be trained in CTD evaluation and care. In this training course, which should be a minimum of 1 to 2 d, there should be included an overview of the CTD problem, techniques of taking the history, techniques of a physical assessment, and techniques of care.

Medical History

A thorough history of the problem is important. There are a number of standardized history forms or a company may develop its own form. It is important that there be sufficient questions to determine the nature of the problem. All too often we see in the nurse's records that the problem is "sore wrist". This does not answer such questions as: "Where is the wrist sore? How long has it been sore? What aggravates it? Are there any other medical conditions, such as arthritis or diabetes? Has the person received medical attention for this? What does the person think caused the sore wrist? Has there been previous episodes? Is the person on medication and treatment for other conditions? What is the nature of the job and the description of job? Have there been any recent changes in the job?" All of these questions need to be answered in order to obtain an adequate history.

Physical Assessment

Nurses can be taught to do physical assessments of the upper extremities. This should include evaluation of range of motion of the cervical spine, shoulder, elbow, wrist, and fingers. It should also include careful palpation and observation of the soft tissues from the neck to the hand. Any evidence of crepitus, heat, atrophy, tenderness, or swelling should be noted. Standard tests as Finkelstein's, Tinel's, and Phalen's should be done. This information should be recorded on a standardized format. It is recommended that diagrams of the upper extremities be used to identify the localization the findings. This is of critical importance because it is necessary to know exactly where the problem occurred. Thus, it is essential for an elbow condition to know whether it is on the medial or lateral surface, in the area of the epicondyle, or other surface area. Standardized forms can be used for this purpose.

CTD Log

Most companies with medical facilities keep a log of employee visits to the medical department. However, it is recommended that either a separate log for CTD cases be kept, or at the very least, they be clearly identified on the master log so as to be able to keep a running tally of the number of these cases. In larger facilities, computerization of medical department activity would be helpful. Information pertaining to department job, treatment, and follow up should be included. This should not be confused with the OSHA log which is for the purpose of providing legally mandated information to OSHA.

In-Plant Treatment

Early care for CTDs includes medication, namely, anti-inflammatory agents, cold or heat, exercises for flexibility and later for strengthening, instructions on biomechanical factors, and evaluation of job conditions. All protocols should be approved and signed by a local plant physician. Criteria for a physician referral should also be established. Generally, if the condition is not improving within approximately two days, referral to physician should be made. If there is improvement, but the condition has not cleared, treatment may be continued for several more days. The OSHA Guidelines for Meatpacking Plants[18] has a medical care flow diagram similar to the approach recommended herein.

Anti-inflammatory Medication

Over the counter, nonprescription, anti-inflammatory medications can be used with good results. Ibuprofen is currently the only nonsteroidal, anti-inflammatory drug other than aspirin or acetaminophen that can be used without prescription. It is recommended that 200 mg tablets four times per day be used.

Cold and Heat

Acute tendinous conditions should be treated with cold as should acute sprains and strains. An ice pack or ice massage applied locally to the tendon in the acute stages is helpful. Also, ice should be used where there is swelling. Ice should generally be used where there may be question of nerve entrapment as at the wrist or around the compartments of the hand.

Where there is muscle spasm, which is recurring or persisting for some days, heat is quite often helpful. Ice or heat may be used prior to gentle stretching exercises. Heat may be applied with a hydroculator pack. Ice or heat when used should be used for approximately 20 min four times per day.

It is helpful if the employee is treated twice during the working day and advised to do treatment twice at home. After several days, in-plant treatment can be discontinued and home treatments continued.

Exercises

Very gentle stretching exercises should be encouraged where there is muscle spasm. Stretching should be in the direction opposite from the force that is applied throughout the day, that is if the job requires frequent flexion and gripping, then stretching should consist of opening the hand and extending the wrist. Once the employee is improved, then he/she should be encouraged to continue flexibility exercises frequently before work and several times during the day. Strengthening exercises, once the acute phase is over, can also be helpful. These may consist of squeezing a nerf ball, or specific muscle strengthening as needed. Qualified physical therapists should be involved in muscle strengthening programs.

Instruction in Biomechanics

Principles of biomechanics are discussed in detail elsewhere in this book. It is quite important that these be discussed with the employee at the time that the condition is reported.

Modified Duty

It may be necessary to modify the work for a temporary period. However, our experience is if these conditions are reported very early, it is frequently possible for the employee to be treated with minimal or no modification of the duty. Modified duty should be temporary only. It should be accompanied by progressive "ramping" up toward regular duty. If there is not satisfactory progress, then consideration should be given to permanent reassignment. An in-plant CTD program will not work well if there are large numbers of employees engaged in "make work" light duty jobs.

Care should be taken in assigning modified duty. We have seen many instances where a person has been reassigned to a similar job using the same muscles and, therefore, has not obtained the necessary rest. One form of a light duty job, when practical, is simply to slow down. This can be very helpful where there is piecework, as for example in sewing. Some companies protect the wages of the employees during this temporary period so that they may decrease their rate without having a financial penalty. Where there is linework, it may be possible to provide some type of relief or slowing by having a utility person relieve for a period of time.

The goal of modified work is to allow the person to continue on a job of meaningful work while at the same time providing some degree of rest to the inflamed muscles or tendons. The nurse, physical therapist, or doctor should play an active role in this assignment. It is very helpful to have on-site discussion with supervisors and to be as specific as possible pertaining to the modification of the work. Simply requesting "light duty" is not helpful. Part II, Chapter 7, provides additional insight on this topic.

Supports, Braces, and Splints

For acute tendonitis, splinting of the tendon is usually necessary. In de Quervain's disease, care should be taken that the thumb is splinted. Wrist supports may be helpful where there is generalized soreness in the region of the wrist. Preferably these should be loose and flexible enough to allow reasonable movement while at the same time providing some support. The use of rigid or semirigid splints to the wrist while the person is working may result in raising the elbows while working. This can cause problems with the elbows and shoulders. In general we believe that wrist splints should be cautiously used with supervision for short periods of time. Splints used at night at the very first signs of numbness or developing carpal tunnel syndrome are often quite helpful. Tennis elbow supports are helpful for epicondylitis.

Recently, a number of manufacturers of wrist supports have advertised these as a preventive for carpal tunnel syndrome. Some companies have purchased these at considerable expense and provided them to all employees working on jobs where there is risk of development of carpal tunnel syndrome. We do not believe that it is advisable to use them as a preventive for the following reasons:

1. They may result in abnormal motions of the elbow and shoulder.
2. Continuous support may reduce strength.
3. It is much more preferable for the individual to learn correct biomechanics and the way to do the job rather than be forced with a splint.

Vitamins

Large doses of vitamin C, E, and B_6 have been advocated as both a preventive and a remedy for CTDs. A number of plants have provided these to employees. As might be expected, there are often glowing testimonials to their benefit. While there may be some rationale in utilizing vitamin C because of its benefit to healing, there has been no controlled study to determine whether vitamins are actually helpful.

There are a number of articles in the medical literature indicating that pyridoxine (vitamin B_6) may be helpful in the treatment of carpal tunnel syndrome (See Part I, Chapter 3). There are others that indicate it is useless. In early cases, it may be helpful, but again, there is no controlled study. If used, it should be used with care and a dosage of 300 mg/d should not be exceeded because of the possibility of neurological damage from the pyridoxine.

The Role of the Physical Therapist

A number of companies have engaged the services of physical therapists, often with considerable benefit. Physical therapists may come to the plant or treat people referred to their office. The physical therapy program should stress active reconditioning and not passive modalities such as ultrasound and heat, except in the very acute stage. Physical therapists also have training in biomechanics and can often be very helpful in educating the employee in correct biomechanics and in helping with placement in modified work. A physical therapist who is experienced with industrial problems is an important member of the CTD program team.

The Role of the Physician

Every plant with a CTD program should have a physician who can approve orders and protocols, evaluate patients with CTDs, help with modified duty assignments, provide consultation to the nurse, and evaluate and treat CTD patients. Selection of this physician is very important. This physician should have the basic knowledge of CTD diagnosis and management. The physician should be oriented toward nonsurgical treatment of these conditions and have a great deal of patience in dealing with them. He/she should use consultants for second opinions, but maintain the management of the case. It is important that the physician selected be carefully oriented into the company's program. This is so irrespective of the level of training and experience of the physician. We recommend an orientation seminar for physicians who will be handling the CTD program in the plant.

Company personnel often ask what type of specialists should be used for CTD referral. No single specialty in and of itself appears to provide superior results. So much depends upon the individual and the philosophy of the physician. Some assume that best results in treatment of CTDs will be obtained by referring to hand surgeons. This is not necessarily true. On the other hand, when surgery is necessary, a qualified hand, orthopedic, plastic, or neurosurgeon is needed. It is important that the physician not be so limited in scope that he can not evaluate and treat conditions arising in the neck, shoulder, elbow, forearm, and hand. Many occupational physicians have developed considerable expertise in handling of CTDs and also have a considerable knowledge of workplace problems. Primary care physicians in family and internal medicine practice may also provide care for these if they are willing to develop the experience and expertise. Neurologists who have interest in these conditions can be of considerable assistance in evaluation and in helping localize the lesions. Physiatrists specialize in physical medicine and rehabilitation. Many of them do electrodiagnostic evaluations and are interested in workplace problems, and some have gained considerable experience and expertise in dealing with them.

Most CTD cases need conservative care. This can be provided by a wide variety of specialties and more important than specialty is the interest, predisposition, and the confidence in which he/she is able to generate in the patient. Obviously when surgical care is indicated, the appropriate surgical specialty is preferred.

Surgical vs. Nonsurgical Treatment of CTDs

In an earlier chapter, differential diagnosis was discussed. In recent years carpal tunnel surgery has become extremely common. However, it has been observed that patients with occupational carpal tunnel syndrome do not do so well after surgery as those with the nonoccupational type.[19] Some surgeons have even recommended against operating upon most occupational cases.

There are a number of reasons why carpal tunnel surgery is not as effective in occupationally related cases.

Frequently, the problem is not carpal tunnel syndrome, but is one of a myofascial pain syndrome some of which cause paresthesias, or proximal entrapments; for example, thoracic outlet syndrome, cervical radiculopathy, or pronator teres syndrome. Even when there are clear-cut indications that the condition is carpal tunnel syndrome, it is often complicated by these other conditions. Without attention to these complicating factors, surgery will fail.

If indeed the carpal tunnel syndrome has been caused by occupation, even though surgery may relieve it, it is prone to re-occur if the occupational factors that caused it originally are not corrected. Although the carpal tunnel is enlarged by the surgery, eventually, continued mechanical stress and swelling may cause recurrence. Many surgeons are now advocating that employees do not return to the same work after surgery. Therefore, the question must be asked: "If a job change is necessary for this individual, should it not be made prior to surgery rather than afterward?" This would then provide the individual with an opportunity to recover without surgery.

There are a great number of psychosocial and socioeconomic factors associated with CTDs. These have been discussed before, and they do indeed complicate the healing process. Without adequately dealing with them, surgery will not succeed.

Sometimes we hear that since conservative treatment has failed, why not try surgery? This can be helpful where there is clear-cut carpal tunnel syndrome with a minimum of complicating factors and this has not responded to conservative treatment. However, when symptoms are vague, indications are unclear, and there are many complicating factors, surgery usually fails or only provides temporary relief. Although carpal tunnel surgery is considered to be relatively simple, it is not without its complications. There could be a be a painful scar, and there is a period of inactivity which requires rehabilitation. Surgery tends to provide, for some, an overly simplistic answer to the problem. Thus, if this is an upper extremity problem involving a number of areas, the patient feels that a cure should be forthcoming from the surgery and is quite disappointed when this doesn't occur. Along with a scar and a period of inactivity, the situation has been made worse. Reflex sympathetic dystrophy may develop in CTD cases whether or not there is surgery, but is a definite complication of carpal tunnel surgery in a small percentage of cases. This condition, when it occurs post surgery, is a devastating experience.

In summary, unless indications are clear, surgery will not only not be helpful, but may be harmful.

CASE MANAGEMENT

All of the above measures may be for naught if the CTD case is not carefully managed. Medical and ergonomic consultants who are very experienced with these problems should be involved in case review at an early stage. In-plant health care and management personnel should work closely with the employee supervisors and attending physicians until the case is resolved. Case managers should make every effort to discourage employees and physicians from premature or unnecessary surgery.

REFERENCES

1. Silverstein, B. A., "The Prevalence of Upper Extremity Cumulative Trauma Disorders in Industry," Doctoral Dissertation, (Ann Arbor: The University of Michigan, 1985).
2. McCormach, R. R., Inman, R. D., Wells, A., Berntsen, C., and Imbus, H. R., "Prevalence of tendinitis and related disorders of the upper extremity in a manufacturing workforce," *J. Rheumat.,* 17(7):958–964 (1990).
3. Bove, F., Litwak, M. S., Arezzo, J. C., and Baker, E. L., "Quantitative sensory testing in occupational medicine," *Seminars in Occupational Medicine,* 1(3):185–189 (1986).
4. Mathiowetz, V., "Grip and pinch strength: normative data for adults," *Arch. Phys. Med. Rehabil.,* 66:69–74 (1985).
5. Feierstein, M. S., "The performance and usefulness of nerve conduction studies in the orthopedic office," *Orthop. Clini. N. Am.,* 19(4):859–866 (1988).
6. Kimura, I. and Ayyar, D. R., "The carpal tunnel syndrome: electrophysiological aspects of 639 symptomatic extremities," *Electromyogr. Clin. Neurophysiol.,* 25:151–164 (1985).
7. Redmond, M. D. and Rivner, M. H. "False positive electrondiagnostic tests in carpal tunnel syndrome," *Muscle and Nerve* 11:511–517 (1988).
8. Dellon, A. L., "The vibrometer," *Plast. Reconstr. Surg.,* 71(3):427–431 (1983).
9. Moody, L., Arezzo, J., and Otto, D., "Evaluation of workers for early Peripheral neuropathy:the role of existing diagnostic tools," *Seminars in Occupational Medicine,* 1(3):153–162 (1986).
10. Katims, J. J., Naviasky, E. H., Rendell, M. S., Lorenz, K. Y., and Bleecker, M. L., "Constant current sine wave transcutaneous nerve stimulation for evaluation of peripheral neuropathy," *Arch. Phys. Med. Rehabil.,* 68:210–213 (1987).
11. Masson, E. A., Veves, A., Fernando, D., and Boulton, A. J. M., "Current perception thresholds: a new, quick, and reproducible method for the assessment of peripheral neuropathy in diabetes mellitus," *Diabetologia,* 32:724–728 (1989).
12. Lundborg, G., Lie-Stenstrom, A, K., Sollerman, C., Stromberg, T., and Pyykko, I., "Digital vibrogram: a new diagnostic tool for sensory testing in compression neuropathy," *J. Hand Surg.,* 11A:693–699 (1986).
13. Arezzo, J. C., Schumburg, H. H., and Laudadio, C., "Thermal sensitivity tester/device for quantitative assessment of thermal sense in diabetic neuropathy," *Diabetes,* 35:590–592 (1986).
14. Gelberman, R. H., Szabo, R. M., Williamson, R. V., and Dimick, M. P., "Sensibility testing in peripheral-nerve compression syndromes," *J. Bone Joint Surg.,* 65-A(5):632–638 (1983).
15. Bell, J. A., "Light touch — deep pressure testing using Semmes-Weinstein monofilaments," *Rehabilitation of the Hand,* (St. Louis: C.V. Mosby, 1990), Ch 43, pp. 585–593.
16. Felton, J. S., *Occupational Medical Management/A Guide to the Organization and Operation of In-Plant Occupational Health Services.* (Boston/Toronto/London: Little, Brown and Company, 1990)

17. American Medical Association, "Guide to Small Plant Occupational Health Programs," *Arch. Environ. Health*, 5:383–392 (1962), Revised 1973.

18. U.S. Department of Labor, Occupational Safety and Health Administration (OSHA) and Bureau of Labor Statistics (BLS), *Ergonomics Program Management Guidelines for Meatpacking Plants*

19. Louis, D. S., "Cumulative trauma disorders," *J. Hand Surg.*, 12A(5)(Part 2):823–825 (1987).

Chapter 7
Minimizing Lost Time

Upper extremity CTDs may follow a similar natural history to that of low back pain. Work-related low back pain has been more extensively studied by rehabilitation and medical experts. The success of various interventions and rehabilitation methods for work-related low back pain provides a model to develop intervention strategies for CTDs. The familiar 20/80 rule applies to the probability that low back pain sufferers will return to work. Within six weeks of injury, 80% of workers with acute low back pain return to work. The remaining 20% of workers will develop chronic low back problems, will not return to work, and represent 80% of the total medical costs related to back pain. Further, if a worker has not returned to work after 6 months, rehabilitative measures become less successful. Study results (Nachemson, 1983) reflect a 30% return-to-work rate after 6 months of injury; 10% return to work after 1 year.[1] These startling statistics emphasize the importance of aggressively addressing problems after their onset, and led to the "work hardening" movement in rehabilitation intervention.

The concept of work hardening borrows from athletic training and focuses on five stages of physical rehabilitation for return to work:

- **Healing process** — depending on the severity of injury, the affected body part is either immobilized or rested. Specific procedures are performed to aid healing and resolve inflammatory processes.

- **Gentle movement** — the joint/body part is moved through its complete range of motion to maintain flexibility and movement *during* the healing process. Gentle movement is actually assistive in the resolution of inflammatory processes and regeneration of musculoskeletal tissues.

- **Progressive resistive exercises** — the muscles are strengthened to recondition the body for physical work demands. This is similar to weight training in football.

- **Work simulation activities** — through simulation, the employee "re-learns" his/her job, focusing on proper body mechanics, coordination, strength, and endurance. This is a self-paced activity performed away from the actual workstation site. These activities are progressive in nature, until the employee demonstrates safe and pain-free performance of job tasks.

- **Graduated return to work process** — the employee returns to the actual workstation and performs paced job tasks. His/her exposure time is progressed until the employee can work an entire shift.

Promoting early movement by minimizing immobilization and rest aids the healing process. This was documented in a study performed by Deyo et al. (1986) of patients with mechanical low back pain.[2] These investigators compared the consequences of 2 d of bed rest for acute low back pain, with those of 7 d bed rest. The group that rested only 2 d missed 4 to 5% fewer days of work than those resting 7 d. No differences in functional outcome was noted between the two groups The negative effect of bed rest is thought to be related to muscle deconditioning and a reduction in vertebral bone density.[3] This study supports a trend towards more aggressive and earlier mobilization of workers with noncomplicated mechanical back pain.

The effects of splinting the wrist can mirror those of placing a worker with low back pain on strict bed rest. With immobilization, no active movement can occur, resulting in rapid muscle deconditioning. The loss of muscle mass, known as atrophy, is the reason for such deconditioning. Strength loss is the most direct consequence of immobilization and atrophy. Muscle strength decreases most dramatically during the first week of immobilization, with less weakening occurring in later weeks.[4] Atrophic muscles suffer from reduced circulation, metabolic changes, and an increase in connective ("scar") tissue that replaces the nonmoving healthy muscle tissue. Due to these changes, the recovery time is often much longer than the total time of immobilization. Thus, 1 week of wrist splinting can result in several weeks of reduced ability to perform at the presplinting work capacity.

In general, it appears that acute musculoskeletal fatigue and inflammation responds *with the least disability* to early efforts to resolve the inflammatory process, accompanied by functional movement. The more severe CTDs, if requiring time away from work for rest or immobilization, also respond positively to aggressive rehabilitation through active movement and job-specific reconditioning.

PRINCIPLES OF RECONDITIONING

Reconditioning involves three principles for optimum acquisition of strength, endurance, coordination, and skill:

Intensity — Strength is obtained when the muscles are exercised at a moderate level of effort (30 to 60% of maximum strength).[4,5]

Duration — Endurance is improved by performing a repetitive task, with sufficient recovery time so that circulation and metabolism replenish the required level of oxygen and fuels to repeat the task. Obviously, if a worker has lost strength, each effort will demand a significant percentage of the remaining strength and more recovery time will be necessary. Initially, a worker will need more frequent breaks until strength is sufficient to perform the repetitive task with less need for recovery time. Repetition leads to skill and coordination, but repetition requires endurance. A worker will be able to progress more rapidly through the duration training, with fewer rest periods, for lighter (<30% maximum voluntary contraction) jobs than for physically demanding jobs.

Specificity — Coordination and skill are based on reaction time and movement time. These are both functions of strength and the nervous system. Movement efficiency evolves from the nervous system's "learning" a motor pattern, so that it requires less processing time to "think through" the task. Body awareness and strength to coordinate all muscles to perform an efficient movement is also necessary.[5] Practice of activity-specific motions enhance motor learning for skill and coordination.

Therefore, work hardening (job-specific reconditioning) requires that many aspects of conditioning take place at once. The progression of muscle function should generally follow this pattern:

- *Light movement and flexibility maneuvers,* to maintain circulation and mobility
- *Strength training,* to prepare for duration and motor pattern learning, to develop coordination, skill, and efficient motor patterns
- *Duration training,* to increase speed and endurance with activities, and finally
- *Skills training* to learn the mechanics of performing the job

Companies that utilize in-plant training programs for new employees can actually perform work hardening at the work site. In using training curves, it is important to remember that each person will progress along the training curve at a different pace. Therefore, incorporating these four training levels, from light activity to actual performance of work, will vary for each new or rehabilitating employee.

Training levels should reflect varying amounts of recovery time throughout each day. Progression to the next training level should be allowed only when the employee demonstrates skill (yield, safety, quality requirements) *and* physical capability (ability to meet strength and repetition requirements at that training level with a minimum of muscle soreness). By requiring that the employee qualify at each training level, both overuse and under-progression will be discouraged. Some employees will rapidly progress on a daily basis, others may require up to 3 to 6 weeks before demonstrating the physical ability and skill to meet 100% performance requirements.

Once the worker begins work simulation activities, a performance expectation may begin at 50%, and progress upward until reaching 100%.[6] This works well if the worker performs a self-paced job, as he can regulate work output during the shift. For example, if prior to injury, a worker generates an average output of 500 units completed in an 8-h shift with two 15-min breaks and one 30-min lunch, the return-to-work output progression is

Hourly rate	=	71.5 units
50%	=	36 units/h
70%	=	50 units/h
85%	=	61 units/h
90%	=	64 units/h
100%	=	71.5 units/h

The 50.3-s cycle required to complete 71.5 units is initially expanded to 100 s. This is then progressively reduced as output increases.

If the worker performs externally paced job tasks, the approach is modified so that actual line time is regulated. One method may be to use the returning worker as an "apprentice", where the worker is only responsible for "every other" unit while sharing a job with another worker. An alternative approach is to limit the worker's exposure to a fully paced line to short bouts of time. With this arrangement, the ratio of on-line and off-line time begins with short on-line exposures and long episodes of off-line activities. Progressively, this relationship reverses, until the worker is able to tolerate an increasing duration of on-line time at line speed performance.

It is important to remember the natural progression of muscular conditioning when designing a "work hardening" training program. Basic motor patterns and job skills related to grasp type, placement of work, yield, quality, and efficient use of motions must be learned first. During the first day of hire, cross training, or return to work, the worker should observe videotapes of good and poor job methods. The worker should practice the job at a stationary work area, away from the paced line, or at his regular work station, if self-paced. The trainer should spend time with the worker to review proper job techniques and to ensure that he performs at his own pace until quality and method skills are intact with a minimum of muscle soreness. Once motor patterns are learned, the worker can progress to on-line or timed performance. During these initial days of skill acquisition and self-paced work, regular breaks should be allowed every hour for alternate activities: job-specific strengthening exercises, stretching, ice/heat to minimize soreness, observation of videotape for self-analysis of work methods, or observations of others' methods. Breaks are not to be perceived as "favor" time to sit in the cafeteria, gossip with friends, disrupt others, or otherwise act as "loafing". If break times are not constructively used and monitored by the trainers, the program can quickly assume a negative image and will not be taken seriously by management or employees.

"Light duty" programs can work . . . if done properly. Many plants are moving toward this "work hardening" approach to rehabilitate workers experiencing work-related injuries and illnesses, as it focuses their return to work on job-specific activities, minimizes nonproductive lost time, and contributes to a positive rehabilitation of the injury/illness. Job specificity is important to meaningful rehabilitation of the returning worker. If a worker were hired off the street to perform a specific assembly job, we would think it ridiculous to have the new employee clean rest rooms, pick up cigarette butts, or perform other menial tasks prior to introducing the new worker to the job for which he was hired. Intuitively, we know that the menial tasks do nothing to prepare the new employee to successfully perform that job. However, many plants employ a light duty program which requires an injured worker to perform menial tasks. These tasks may contribute nothing to the plant's productivity, yet often increase overall labor costs. Once this employee exhibits that he can work pain-free in these menial tasks, the doctor will release him for full duty in his regular job. The employer will then expect the worker to immediately perform at pre-injury levels, regardless of the interruption of work performance and the associated deconditioning and skill loss. Employers are surprised and angry when the worker's symptoms resurface. Often, at this point the worker is labelled a "malingerer".

Many employers incur substantial costs associated with lost time simply because they refuse a worker's return unless 100% capable of pre-injury performance. No productivity is gained for the expense. If the worker is allowed to return and perform at progressive levels on a work hardening program, productive work can be exchanged for the cost incurred. However, this exchange will require close coordination between plant management, supervisor, trainer, and physician, so that the worker is not over- or underperforming. With this effort, lost time can be minimized, the cost of rehabilitation can be offset by some level of productivity, and the odds for successful return to work are improved for the worker.

With these concepts in mind, it is proposed that "light duty" be removed from the work rehabilitation system altogether. Light duty concepts should be replaced with work hardening concepts. In work hardening, the worker's previous job can be made "lighter" in output expectation, progressing into greater output expectations until a full performance level is achieved. Additionally, prior to a worker's return, an ergonomic evaluation of the equipment, workstation, and methods should be reviewed. If modifications can be made to the method, layout, tools, etc., then that modification itself represents a replacement of the former job with a "lighter" job. Ergonomic modifications should also quicken the worker's ability to return, as the physical demands are lessened.

Occasionally, the worker is capable of returning to the workplace, but is restricted from his previous job in order to allow the injured body part to heal. In this case, alternate duty may be pursued. Alternate duty will differ for each employee and each specific injury or illness, and requires close communication with the physician and/or physical therapist to quantify the worker's capabilities. For example, if a worker experiences de Quervain's disease (tenosynovitis of thumb extensor tendons), alternate duty would involve a job requiring the use of that hand, but excluding pinching or specific use of the thumb. If one finger on the hand is affected, the remaining four fingers do not also need to lay idle! Such a restriction would require the worker to use only one hand and arm, and would severely limit the availability of alternate duty jobs while weakening the remaining healthy tissue.

Rather, jobs should be evaluated as to whether that specific injured body part will be unduly stressed as to prevent healing. For example, in a chicken deboning operation, an employee suffered tendonitis in the dominant shoulder. The physician sent the employee back to work with a restriction that read "no use of hand tools". This restricted the use of scissors, small power tools, and knives, even if used in the lightest and most neutral upper body position. Therefore, the supervisor had extreme difficulty locating an alternate duty job that allowed the employee to work without hand tools. A more effective alternate duty restriction would have read: "No lifting, reaching, pushing or pulling with the dominant shoulder greater than 30 degrees away from the body." If the physical limitation is well defined by the physician, the plant and management can be more accommodating when locating alternate duty jobs. The key is communication and respect for both the management's concerns and the proper rehabilitation of the injury in question.

To accommodate injury-specific restrictions, each job must be evaluated and ranked according to exertion level for each particular body part. This information can be stored in a data base, so that when a worker experiences a CTD and must be restricted, the jobs available to that worker can be reviewed to determine which offer the least stress on the affected body part. Those jobs can then be videotaped for the physician's review, to select an alternate job that best fits the worker's capabilities and restrictions. Remember — the more specific the physician *and* the company can be regarding work exertion levels and employee physical capability *per body part*, the sooner the worker can return to productive work.

REFERENCES

1. Nachemson, A., "Work for all:for those with low back pain as well," *Clinical Orthopedics and Related Research*, 179:77–85 (1983).

2. Deyo, R. A., Diehl, A. K., and Rosenthal, M., "How many days of bed rest for acute low back pain?" *New Eng. J. Med.*, 315(17):1064–1070 (1986).

3. Bigos, S. J. and Battiè, M. C., "Acute Care to Prevent Back Disability," *Clin. Orthop.*, 221:121–130 (1987).

4. Appell, H. J., "Muscular atrophy following immobilization. A review," *Sports Med.*, 10(1):42–58 (1990).

5. Sharkey, B. J., *Physiology of Fitness* (Champagne, IL: Human Kinetic Publishers, 1979), p. 73.

6. Day, D. E., "Preventive and return to work aspects of cumulative trauma disorders in the workplace," *Seminars in Occupational Medicine*, 2(1):57–63 (1987).

Index